METHODS IN MOLECULAR BIOLOGY™

Series Editor
John M. Walker
School of Life Sciences
University of Hertfordshire
Hatfield, Hertfordshire, AL10 9AB, UK

For other titles published in this series, go to
www.springer.com/series/7651

Stem Cells for Myocardial Regeneration

Methods and Protocols

Edited by

Randall J. Lee

Department of Medicine, Cardiovascular Research Institute, and Institute of Regeneration Medicine, University of California San Francisco, San Francisco, CA, USA

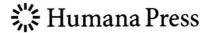

Editor
Randall J. Lee, MD, Ph.D.
Department of Medicine,
Cardiovascular Research Institute, and
Institute of Regeneration Medicine,
University of California San Francisco,
San Francisco, CA
lee@medicine.ucsf.edu

ISSN 1064-3745   e-ISSN 1940-6029
ISBN 978-1-60761-704-4   e-ISBN 978-1-60761-705-1
DOI 10.1007/978-1-60761-705-1
Springer New York Dordrecht Heidelberg London

Library of Congress Control Number: 2010931703

Preface

The field of regenerative medicine is in its infancy state. Enthusiasm for the potential of organ regeneration lies with the potential pluripotency of stem cells to differentiate into various tissue types. This volume of Methods in Molecular Biology will focus on the use of stem cells for myocardial repair and regeneration. The emphasis of this issue will be to provide basic scientists, translational investigators, and cardiologists a means to evaluate the efficacy and safety of stem cells in a standardized fashion for myocardial regeneration.

Many different cell types have been considered for myocardial repair. Adult cardiomyocytes are unable to survive even when transplanted into normal myocardium. The use of fetal or neonatal cardiomyocytes is not a feasible source of cells due to ethical concerns and donor availability. Therefore, the use of pluripotent stem cells has become the focus of a cell source for myocardial repair and regeneration. A variety of stem cell types have been suggested to participate in myocardial repair. This has led the investigators to search for the "optimal cell type for myocardial repair". Reliable isolation of the cell source with the ability to expand the cell population is a prerequisite. In the first section of this book, methods for isolation of commonly used stem cells being investigated for myocardial regeneration are presented.

Once a stem cell source has been selected, the stem cell needs to be tested in an appropriate animal model before being translated into clinical practice. Section 2 discusses both rodent and large animal models. The pros and cons of utilizing each of the models are discussed, as well as obtaining consistent myocardial pathology to test whether the stem cells improve function. Techniques used to assess left ventricular function are described for both rodent and large animals, as well as methods to identify stem cells and their effect on myocardial repair.

Understanding the developmental process of the human heart is paramount to developing strategies for myocardial regeneration. Knowledge of the cellular components of the heart and their response to injury is crucial in designing experiments and therapies for myocardial repair and regeneration. Discrepancies in results of stem cell differentiation into cardiomyocytes and its efficacy are commonly dependent on the interpretation of the histological results. Section 3 reviews the histological characteristics of the developing and normal myocardium and provides the histological chronology of the heart following a myocardial infarction. Strategies for myocardial regeneration also include means to develop a functional vascular system. It is important to discriminate between increases in capillary density that commonly do not increase blood flow and arteriogenesis that will lead to an increase in blood flow. A detailed analysis of angiogenesis and methods to delineate the types of vasculature produced by stem cells are also discussed in section 3.

Once a stem cell is transplanted into the myocardium, it is of great importance to determine its fate and to assure safety. MRI and molecular imaging enable the identification and tracking of transplanted stem cells. The use of superparamagnetic iron oxides to label stem cells has enabled investigators to utilize MRI to assess the injection of stem cells into the injured area and its effect on both segmental and global left ventricular function and myocardial perfusion. Transfection of stem cells with a reporter gene allows the reporter probe to produce a signal detectable by commonly used imaging modalities. Molecular genetic imaging is confined to viable stem cells and the population of stem cells transfected, thus allowing for longitudinal tracking of stem cells. Molecular imaging has been particularly useful in following embryonic stem cells and their propensity to form teratomas. Recently, the beneficial effects of autologous stem cell therapy have been attributed to paracrine effects. The use of a genetic fate-mapping approach is reviewed in section 4 to study adult cardiomyocyte replenishment following an injury. The use of the tools in section 4 will allow investigators to address challenges of stem cell therapy such as stem cell retention, engraftment, and safety, and investigate the mechanisms of stem cell therapy.

The emphasis of myocardial regeneration has focused on improvements of left ventricular function; however, an electrically integrated transplanted stem cell with its surrounding environment is necessary to mitigate abnormal arrhythmias and optimize electromechanical performance. Both in vitro assessment of cellular electrophysiological properties and cell-to-cell communication can be accomplished with multielectrode array recordings and optical mapping. These studies can be complemented with either ex vivo optical mapping or in vivo electrophysiology studies. These methods are presented in section 5.

Tissue engineering techniques have been used to enhance cell retention and create the microenvironment to allow for stem cell survival. More recently, the extracellular matrix or functional groups derived from extracellular matrix proteins have been shown to influence stem cell binding, the production of growth factors by the stem cell, and stem cell differentiation. In the final section of this volume, a strategy for investigating the effects of the extracellular matrix on stem cell renewal and differentiation is presented.

The methods presented in this volume of Methods in Molecular Biology attempt to highlight techniques and strategies to be utilized in investigating the many challenges that need to be addressed before stem cell therapy can become a mainstream therapy for myocardial regeneration.

San Francisco, CA   *Randall J. Lee*

Contents

Contributors

GIL ARBEL • *The Bruce Rappaport Faculty of Medicine, Sohnis Laboratory for Cardiac Electrophysiology and Regenerative Medicine, Technion-Israel Institute of Technology, Haifa, Israel*

ANDREW J. BOYLE, MBBS, PHD • *Cardiology Division, Department of Medicine, University of California, San Francisco, CA, USA*

NIRMA BUSTAMANTE • *Department of Medicine, University of Texas Southwestern Medical Center at Dallas, Dallas, TX, USA*

NAIMA CARTER-MONROE, MD • *CVPath Institute, Gaithersburg, MD, USA*

OREN CASPI • *The Bruce Rappaport Faculty of Medicine, Sohnis Laboratory for Cardiac Electrophysiology and Regenerative Medicine, Technion-Israel Institute of Technology, Haifa, Israel*

SUNNY S.-K. CHAN • *Graduate Institute of Clinical Medicine and Research Center for Clinical Medicine, National Cheng Kung University & Hospital, Tainan City, Taiwan*

JYH-HONG CHEN • *Department of Medicine, National Cheng Kung University, Tainan City, Taiwan*

KAREN L. CHRISTMAN, PHD • *Department of Bioengineering, University of California, San Diego, CA, USA*

WANGDE DAI, MD • *Good Samaritan Hospital, The Heart Institute, Division of Cardiovascular Medicine of the Keck School of Medicine at University of Southern California, Los Angeles, CA, USA*

CHUNHUA DING, MD, PHD • *Cardiac Electrophysiology Section of Cardiology, Department of Medicine, University of California, San Francisco, CA, USA*

THOMAS H. EVERETT IV, PHD • *Cardiac Electrophysiology Section of Cardiology, Department of Medicine, University of California, San Francisco, CA, USA*

AMIRA GEPSTEIN, PHD • *The Bruce Rappaport Faculty of Medicine, Sohnis Laboratory for Cardiac Electrophysiology and Regenerative Medicine, Technion-Israel Institute of Technology, Haifa, Israel*

LIOR GEPSTEIN, PHD • *The Bruce Rappaport Faculty of Medicine, Sohnis Laboratory for Cardiac Electrophysiology and Regenerative Medicine, Technion-Israel Institute of Technology, Haifa, Israel*

NICHOLAS GRECO, PHD • *Division of Hematology/Oncology, Department of Medicine, School of Medicine, Case Western Reserve University, Cleveland, OH, USA*

JOSHUA M. HARE, MD • *Interdisciplinary Stem Cell Institute and Division of Cardiology, University of Miami Miller School of Medicine, Miami, FL, USA*

PATRICK C. H. HSIEH • *Graduate Institute of Clinical Medicine and Research Center for Clinical Medicine, National Cheng Kung University & Hospital, Tainan City, Taiwan*

IRIT HUBER, PhD • *The Bruce Rappaport Faculty of Medicine,
Sohnis Laboratory for Cardiac Electrophysiology and Regenerative Medicine,
Technion-Israel Institute of Technology, Haifa, Israel*

MASAAKI II, MD, PhD • *Group of Vascular Regeneration Research,
Institute of Biomedical Research and Innovation, Minatojima-minamimachi,
Chuo-ku, Kobe, Japan*

DOROTA A. KEDZIOREK, MD • *Russell H. Morgan Department of Radiology
and Radiological Science, The Johns Hopkins University, Baltimore, MD, USA*

ROBERT A. KLONER, MD, PhD • *Good Samaritan Hospital, The Heart Institute,
Division of Cardiovascular Medicine of the Keck School of Medicine
at University of Southern California, Los Angeles, CA, USA*

FRANK D. KOLODGIE, PhD • *CVPath Institute, Gaithersburg, MD, USA*

DARA L. KRAITCHMAN, VMD, PhD • *Russell H. Morgan Department of Radiology
and Radiological Science, The Johns Hopkins University, Baltimore, MD, USA*

ELENA LADICH, MD • *CVPath Institute, Gaithersburg, MD, USA*

MARY J. LAUGHLIN, MD • *Division of Hematology/Oncology,
Department of Medicine, School of Medicine, Case Western Reserve University,
Cleveland, OH, USA*

RANDALL J. LEE, MD, PhD • *Department of Medicine, Cardiovascular
Research Institute, and Institute of Regeneration Medicine,
University of California San Francisco, San Francisco, CA, USA*

RONGLIH LIAO • *Cardiac Muscle Research Laboratory, Cardiovascular Division,
Department of Medicine, Brigham and Women's Hospital,
Harvard Medical School, Boston, MA, USA*

ELIZABETH A. LIPKE, PhD • *Department of Chemical Engineering,
Auburn University, Auburn, AL, USA*

HEATHER D. MAYNARD, PhD • *Department of Chemistry & Biochemistry,
University of California, Los Angeles, CA, USA*

IAN K. MCNIECE, PhD • *Interdisciplinary Stem Cell Institute
and Division of Cardiology, University of Miami Miller School of Medicine,
Miami, FL, USA*

ANGELOS OIKONOMOPOULOS • *Cardiac Muscle Research Laboratory,
Cardiovascular Division, Department of Medicine, Brigham and Women's
Hospital, Harvard Medical School, Boston, MA, USA*

OTMAR PFISTER • *Myocardial Research, Department of Biomedicine,
and Division of Cardiology, University Hospital Basel, Basel, Switzerland*

SHARAD RASTOGI, MD, MS • *Division of Cardiovascular Medicine,
Department of Internal Medicine, Henry Ford Hospital,
Henry Ford Heart & Vascular Institute, Detroit, MI, USA*

KARIM SALLAM, MD • *Division of Cardiology, Department of Medicine,
Stanford University School of Medicine, Stanford, CA, USA*

KONSTANTINA-IOANNA SERETI • *Cardiac Muscle Research Laboratory,
Cardiovascular Division, Department of Medicine, Brigham and Women's
Hospital, Harvard Medical School, Boston, MA, USA*

YING-ZHANG SHUEH • *Graduate Institute of Clinical Medicine and Research Center for Clinical Medicine, National Cheng Kung University & Hospital, Tainan City, Taiwan*

MATTHEW L. SPRINGER, PHD • *Department of Medicine, Cardiovascular Research Institute and Institute of Regeneration Medicine, University of California, San Francisco, CA, USA*

SHIH-JUNG TSAI • *Nano-powder & Thin Film Technology Center, New Industry Creation Hatchery Center, Industrial Technology Research Institute, Tainan, Taiwan*

LESLIE TUNG, PHD • *Department of Biomedical Engineering, The Johns Hopkins University, Baltimore, MD, USA*

RENU VIRMANI, MD • *CVPath Institute, Gaithersburg, MD, USA*

MICHAL WEILER-SAGIE • *The Bruce Rappaport Faculty of Medicine, Sohnis Laboratory for Cardiac Electrophysiology and Regenerative Medicine, Technion-Israel Institute of Technology, Haifa, Israel*

SETH WEINBERG, B.S • *Department of Biomedical Engineering, The Johns Hopkins University, Baltimore, MD, USA*

HUA-LIN WU, PHD • *Cardiovascular Research Center and Graduate Institute of Biochemistry and Molecular Biology, National Cheng Kung University, Tainan City, Taiwan*

JOSEPH C. WU, MD, PHD • *Division of Cardiology and The Department of Radiology and Molecular Imaging Program, Department of Medicine, Stanford University School of Medicine, Stanford, CA, USA*

Chapter 1

Stem Cells for Myocardial Repair and Regeneration: Where Are We Today?

Randall J. Lee

Abstract

An overview for the use of stem cells for myocardial repair and regeneration is provided. The overview provides the rationale for use of stem cells in myocardial repair. Potential stem cell types and technological challenges are highlighted.

Key words: Stem cells, Myocardial repair, Tissue engineering

Cardiovascular disease (CVD) remains the leading cause of morbidity and mortality worldwide and is predicted to be the leading cause of death by 2020 (1, 2). A large proportion of cardiovascular disease results from ischemic heart disease. Annually, over 19 million people worldwide sustain a cardiac event, with approximately one million occurring in the United States (3). In the United States, 60 million patients suffer cardiovascular disease costing the healthcare system approximately $186 billion annually (4, 5). Approximately two-thirds of patients surviving a myocardial infarction are left with debilitating congestive heart failure. Despite the advances in interventional procedures and medical treatment to reduce mortality in patients with ischemic heart disease, the number of patients with refractory myocardial ischemia and congestive heart failure is rapidly increasing (6, 7). For end-stage heart failure, heart transplantation is the only viable treatment. However, the number of heart transplantations is limited by the availability of donor hearts (8, 9).

Until recently, the notion that the myocardium is unable to regenerate following an injury was commonly accepted (10, 11). Adult cardiomyocytes were thought to be terminally differentiated and unable to replicate. Therefore, following a myocardial infarction (MI), the death of cardiomyocytes leads to scar tissue

Randall J. Lee (ed.), *Stem Cells for Myocardial Regeneration: Methods and Protocols*, Methods in Molecular Biology, vol. 660, DOI 10.1007/978-1-60761-705-1_1, © Springer Science+Business Media, LLC 2010

formation and aneurismal thinning of the left ventricle. The death of the cardiomyocytes results in negative left ventricular (LV) remodeling which leads to increased wall stress in the remaining viable myocardium. This process results in a sequence of molecular, cellular, and physiological responses that lead to LV dilatation and progressive heart failure. Although the exact mechanisms of heart failure are unknown, it is suggested that LV remodeling may contribute independently to its progression (12).

The discovery of cardiomyocyte replication in animal and human myocardium has rejuvenated enthusiasm for the field of regenerative cardiology. In a series of elegant experiments, Anversa and colleagues have described cardiac stem cells and their differentiation into cardiomyocytes (13–16). Other investigators have described cell cycle markers in normal appearing adult cardiomyocytes suggesting that these cardiomyocytes are undergoing cell division (17). Circulating progenitor cells, perhaps arising from bone marrow, have been suggested to contribute to cardiomyocyte proliferation (18, 19). However, there is controversy involving the origin of the cardiomyocytes, whether they arise from dividing cardiomyocytes, circulating progenitor cells, or resident cardiac stem cells (15, 16, 20, 21).

Cell transplantation therapy has generated promise for repairing damaged myocardium. Despite the controversy in mechanism, preclinical and clinical studies have shown beneficial effects of stem cell therapy after myocardial infarction (MI). Both academic and commercial institutions have demonstrated the feasibility of harvesting autologous stem cells from cardiac patients (22–28). Several independent studies have demonstrated the clinical safety and feasibility of autologous bone marrow mononuclear cells (BMC) to induce myocardial regeneration and neovascularization in patients with ischemic myocardial injury (29–33). Numerous published studies indicate that infused SC promote myocardial repair in both the acute and chronic settings (22, 28, 34, 35) in various measurements of LV function in the 4–14% range are commonly derived from BMC injections even though the majority of cells are rapidly cleared from the heart. However, not all studies have been uniformly beneficial (31–33, 36, 37) and there is significant room to improve the long-term performance of ischemic myocardium in light of the results of the BOOST trial (31). The 1-year follow-up data from the REPAIR-AMI study demonstrating sustained improvement in LV function has become a pivotal trial in validating the use of stem cells for myocardial repair (36, 38). Currently, multiple studies are going on in the US for refractory angina, acute MI, and heart failure.

Despite the promising advances for the use of stem cells for treating heart disease, several challenges of cell therapy for myocardial repair exists. Questions remain as to which cell type is most effective in myocardial repair as well as the optimal time for stem cell transplantation (39, 40). Reliability of cell isolation to obtain a pure cell population and expansion of cells to procure sufficient numbers of cells for transplantation are possible with bone marrow-derived stem cells. However, reliable isolation, expansion, and differentiation of pluripotent stem cells to cardiomyocytes with either human embryonic stem cells (hESC) or induced pluripotent stem cells (iPSC) are being developed. Additionally, limitations of stem cell therapy including delivery, engraftment, electrical integration, safety, and efficacy need to be addressed.

Tissue engineering approaches integrating the use of growth factors and biomaterial scaffolds have been used to enhance cellular transplantation for repair of lost or damaged tissue. In vitro tissue-engineered cardiac constructs (41–56) have been produced, as well as implanting cells on the surface of the myocardium in a polymer scaffold (46, 47, 50, 52, 57–59). An in situ approach of cardiac tissue engineering combining the disciplines of cell biology, material science and engineering principles have been used to repair both acute and chronically damaged myocardium (60, 61). More recently, surface modification of polymers has been shown to induce stem cell production of growth factors and act synergistically with stem cells in producing angiogenesis. The ultimate solution for myocardial repair and regeneration may be the optimization of the combination of stem cells, growth factors, and scaffolds (62).

In this volume of Methods in Molecular Biology, methods and reviews are provided to help investigators assess the potential of stem cells for myocardial repair and regeneration. Guidelines for reliable isolation of commonly assessed stem cells for myocardial repair are presented, as well as methods to determine engraftment, efficacy, electrical integration, and safety.

References

1. He J., Gu D., Wu X., Reynolds K., Duan X., Yao C., Wang J., Chen C. S., Chen J., Wildman R. P., Klag M. J., and Whelton P. K. (2005) Major causes of death among men and women in China. *N Engl J Med* 353,1124–1134.

2. Thom T., Haase N., Rosamond W., Howard V.J., Rumsfeld J., Manolio T., Zheng Z. J., Flegal K., O'Donnell C., Kittner S., Lloyd-Jones D., Goff D. C. Jr., Hong Y., Adams R., Friday G., Furie K., Gorelick P., Kissela B., Marler J., Meigs J., Roger V., Sidney S., Sorlie P., Steinberger J., Wasserthiel-Smoller S., Wilson M., Wolf P. (2006) American Heart Association Statistics Committee and Stroke Statistics Subcommittee. *Circulation* **113**, e85–151.

3. Naghavi M., Libby P., Falk E., Casscells S.W., Litovsky S., Rumberger J., Badimon J. J., Stefanadis C., Moreno P., Pasterkamp G., Fayad Z., Stone P. H., Waxman S., Raggi P., Madjid M., Zarrabi A., Burke A., Yuan C., Fitzgerald P. J., Siscovick D. S., de Korte C. L, Aikawa M., Juhani Airaksinen K. E., Assmann G., Becker C. R., Chesebro J. H., Farb A., Galis Z. S., Jackson C., Jang I. K., Koenig W., Lodder R. A., March K., Demirovic J., Navab M., Priori S. G., Rekhter M. D., Bahr R., Grundy S. M., Mehran R., Colombo A., Boerwinkle E., Ballantyne C., Insull W. Jr., Schwartz R. S., Vogel R., Serruys P. W., Hansson G. K., Faxon D. P., Kaul S., Drexler H., Greenland P., Muller J. E., Virmani R.,

Ridker P. M., Zipes D. P., Shah P. K., Willerson J. T. (2003) From vulnerable plaque to vulnerable patient. A call for new definitions and risk assessment strategies: Part I. *Circulation* **108**, 1664–72.

4. Cohn J. N., Bristow M. R., Chien K. R., Colucci W. S., Frazier O. H., Leinwand L. A., Lorell B. H., Moss A. J., Sonnenblick E. H., Walsh R. A., Mockrin S. C., and Reinlib L. (1997) Report of the National Heart, Lung, and Blood Institute Special Emphasis Panel on Heart Failure Research. *Circulation* **95**, 766–70.

5. Lenfant C. (1997) Fixing the failing heart. *Circulation* **95**, 771–2.

6. Mannheimer C., Camici P., Chester M. R., Collins A., DeJongste M., Eliasson T., Follath F., Hellemans I., Herlitz J., Lüscher T., Pasic M., and Thelle D. (2002) The problem of chronic refractory angina; report from the ESC Joint Study Group on the Treatment of Refractory Angina. *Eur Heart J* **23**, 355–70.

7. Mathur A., and Martin J. F. (2004) Stem cells and repair of the heart. *Lancet* **364**, 183–92.

8. El Oakley R. M., Yonan N. A., Simpson B. M., and Deiraniya A. K. (1996) Extended criteria for cardiac allograft donors: a consensus study. *J Heart Lung Transplant* **15**, 255–9.

9. Keck B., Bennett L., Rosendale J., Daily O., Novick R., and Hosenpud J. (1999) Worldwide thoracic organ transplantation: a report from the UNOS/ISHLT International Registry for Thoracic Organ Transplantation. *Clin Transpl* 35–49.

10. Rumyantsev P. P. (1977) Interrelations of the proliferation and differentiation processes during cardiact myogenesis and regeneration. *Int Rev Cytol* **51**, 186–273.

11. Dorfman J., and Kao R. L. (1997) Myocardial growth and regeneration overview. In: Chiu RC, ed. Cellular cardiomyoplasty: myocardial repair with cell implantation. Austin: Landes Bioscience, 1–25.

12. Mann D. L. (1999) Mechanisms and models in heart failure: A combinatorial approach. *Circulation* **100**, 999–1008.

13. Beltrami A. P., Barlucchi L., Torella D., Baker M., Limana F., Chimenti S., Kasahara H., Rota M., Musso E., Urbanek K., Leri A., Kajstura J., Nadal-Ginard B., and Anversa P. (2003) Adult cardiac stem cells are multipotent and support myocardial regeneration. *Cell* **114**, 763–76.

14. Urbanek K., Torella D., Sheikh F., Angelis A. D., Nurzynska D., Silvestri F., Beltrami C. A., Bussani R., Beltrami A. P., Quaini F., Bolli R., Leri A., Kajstura J., and Anversa P. (2005) Myocardial regeneration by activation of multipotent cardiac stem cells in ischemic heart failure. *Proc Natl Acad Sci U S A* **102**, 8692–7.

15. Nadal-Ginard B., Kajstura J., Leri A., and Anversa P. (2003) Myocyte death, growth, and regeneration in cardiac hypertrophy and failure. *Circ Res* **92**, 139–50.

16. Anversa P., Leri A., and Kajstura J. (2006) Cardiac regeneration. *J Am Coll Cardiol* **47**, 1769–76.

17. Schuster M. D., Kocher A. A., Seki T., Martens T. P., Xiang G., Homma S., and Itescu S. (2004) Myocardial neovascularization by bone marrow angioblasts results in cardiomyocyte regeneration. *Am J Physiol Heart Circ Physiol* **287**, H525–32.

18. Orlic D., Kajstura J., Chimenti S., Limana F., Jakoniuk I., Quaini F., Nadal-Ginard B., Bodine D. M., Leri A., and Anversa P. (2001) Mobilized bone marrow cells repair the infarcted heart, improving function and survival. *Proc Natl Acad Sci U S A* **98**, 10344–9.

19. Jackson K. A., Majka S. M., Wang H., Pocius J., Hartley C. J., Majesky M. W., Entman M. L., Michael L. H., Hirschi K. K., and Goodell M. A. (2001) Regeneration of ischemic cardiac muscle and vascular endothelium by adult stem cells. *J Clin Invest* **107**, 1395–402.

20. Murry C. E., Reinecke H., and Pabon L. M. (2006) Regeneration gaps: observations on stem cells and cardiac repair. *J Am Coll Cardiol* **47**, 1777–85.

21. Chien K. R. (2004) Stem cells: Lost in translation. *Nature* **428**, 607.

22. Patel A. N., Geffner L., Vina R. F., Saslavsky J., Urschel H. C., Kormos R. and Benetti F. (2005) Surgical treatment for congestive heart failure with autologous adult stem cell transplantation: A prospective randomized study. *J Thorac Cardiovasc Surg* **130**, 1631–8.

23. Ince H., Valgimigli M., Petzsch M., de Lezo J. S., Kuethe F., Dunkelmann S., Biondi-Zoccai G., and Nienaber C. A. (2008) Cardiovascular events and re-stenosis following administration of G-CSF in acute myocardial infarction: systematic review and meta-analysis. *Heart* **94**, 610–6.

24. Zohlnhöfer D., Ott I., Mehilli J., Schömig K., Michalk F., Ibrahim T., Meisetschläger G., von Wedel J., Bollwein H., Seyfarth M., Dirschinger J., Schmitt C., Schwaiger M., Kastrati A., Schömig A. (2006) Stem cell mobilization by granulocyte colony-stimulating factor in patients with acute myocardial infarction. *JAMA* **295**, 1003–10

25. Ripa R. S., Jørgensen E., Wang Y., Thune J. J., Nilsson J. C., Søndergaard L., Johnsen H. E., Køber L., Grande P., and Kastrup J. (2006) Stem cell mobilization induced by subcutaneous granulocyte-colony stimulating factor to improve cardiac regeneration after acute ST-elevation myocardial infarction. *Circulation* 113, 1983–92.

26. Valgimigli M., Rigolin G. M., Cittanti C., Malagutti P., Curello S., Percoco G., Bugli A. M., Della Porta M., Bragotti L. Z., Ansani L., Mauro E., Lanfranchi A., Giganti M., Feggi L., Castoldi G., and Ferrari R. (2005) Use of granulocyte-colony stimulating factor during acute myocardial infarction to enhance bone marrow stem cell mobilization in humans: clinical and angiographic safety profile. *Eur Heart J* 26, 1838–45.

27. Baxterpress release (2006) http://www.baxter.com/about_baxter/news_room/news_releases/03-07-06 stem_cell_trial.html.

28. Sinha S., Poh K. K., Sodano D., Flanagan J., Ouilette C., Kearney M., Heyd L., Wollins J, Losordo D., and Weinstein R. (2006) Safety and efficacy of peripheral blood progenitor cell mobilization and collection in patients with advanced coronary heart disease. *J Clin Apher* 21, 116–20.

29. Strauer B. E., Brehm M., Zeus T, Kostering M., Hernandez A., Sorg R. V., Kogler G., and Wernet P. (2002) Repair of infarcted myocardium by autologous intracoronary mononuclear bone marrow cell transplantation in humans. *Circulation* 106, 1913–8.

30. Tse H. F., Kwong Y. L., Chan J. K., Lo G., Ho C. L., and Lau C. P. (2003) Angiogenesis in ischaemic myocardium by intramyocardial autologous bone marrow mononuclear cell implantation. *Lancet* 361, 47–9.

31. Meyer G. P., Wollert K. C., Lotz J., Stgeffens J., Lippolt P., Fichtner S., Hecker H., Schaefer A., Arseniev L., Hertenstein B., Ganser A., and Drexler H. (2006) Intracoronary bone marrow cell transfer after eighteen months' follow-up data from the randomized, controlled BOOST (BOne marrow transfer to enhance ST-elevation infarct regeneration) trial. *Circulation* 113,1287–94.

32. Assmus B., Honold J., Schächinger V., Britten M. B., Fischer-Rasokat U., Lehmann R., Teupe C., Pistorius K., Martin H., Abolmaali N. D., Tonn T., Dimmeler S., and Zeiher A. M. (2006) Transcoronary transplantation of progenitor cells after myocardial infarction. *N Engl J Med* 355, 1222–32.

33. Schachinger V., Erbs S., Elsasser A., Haberbosch W., Hambrecht R., Holschermann H., Yu J., Corti R., Mathey D. G., Hamm C. W., Suselbeck T., Assmus B., Tonn T., Dimmeler S., and Zeiher A. M. (2006) Intracoronary bone marrow-derived progenitor cells in acute myocardial infarction. *N Engl J Med* 355, 1210–21.

34. Archundia A., Aceves J. L., Lopez-Hernandez M., Alvarado M., Rodriguez E., Diaz Quiroz G., Paez A., Rojas F. M., and Montano L. F. (2005) Direct cardiac injection of G-CSF mobilized bone-marrow stem-cells improves ventricular function in old myocardial infarction. *Life Sci.* 78, 279–83.

35. Strauer B. E., Brehm M., Zeus T., Bartsch T., Schannwell C., Antke C., Sorg R. V., Kogler G., Wernet P., Muller H. W., and Kostering M. (2005) Regeneration of human infarcted heart muscle by intracoronary autologous bone marrow cell transplantation in chronic coronary artery disease: the IACT Study. *J Am Coll Cardiol* 46, 1651–8.

36. Schachinger V., Erbs S., Elsasser A., Haberbosch W., Hambrecht R., Holschermann H., Yu J., Corti R., Mathey D. G., Hamm C. W., Mark B., Assmus B., Tonn T., Dimmeler S. and Zeiher A. M. (2006) Improved clinical outcome after intracoronary administration of bone-marrow-derived progenitor cells in acute myocardial infarction: final 1-year results of the REPAIR-AMI trial. *Eur Heart J* 27, 2775–83.

37. Janssens S., Dubois C., Bogaert J., Theunissen K., Deroose C., Desmet W., Kalantzi M., Herbots L., Sinnaeve P., Dens J., Maertens J., Rademakers F., Dymarkowski S., Gheysens O., Van Cleemput J., Bormans G., Nuyts J., Belmans A., Mortelmans L., Boogaerts M., Van de Werf F. (2006) Autologous bone marrow-derived stem-cell transfer in patients with ST-segment elevation myocardial infarction: double-blind, randomised controlled trial. *Lancet* 367, 113–21.

38. Dill T., Schächinger V., Rolf A., Möllmann S., Thiele H., Tillmanns H., Assmus B., Dimmeler S., Zeiher A. M., Hamm C. (2009). Intracoronary administration of bone marrow-derived progenitor cells improves left ventricular function in patients at risk for adverse remodeling after acute ST-segment elevation myocardial infarction: results of the Reinfusion of Enriched Progenitor cells And Infarct Remodeling in Acute Myocardial Infarction study (REPAIR-AMI) cardiac magnetic resonance imaging substudy. *Am Heart J* 157 (3), 541–7.

39. Almsherqi Z., and El Oakley R. (2003) Bone marrow-derived cell transplantation for acute myocardial ischemia. *Circulation* 107, e86–7.

40. Liao R., Pfister O., Jain M., and Mouquet F. (2007) The bone marrow – cardiac axis of myocardial regeneration. *Prog Cardiovasc Dis* **50**, 18–30.

41. Akins R. E., Boyce R. A., Madonna M. L., Schroedl N. A., Gonda S. R., McLaughlin T. A., and Hartzell C. R. (1999) Cardiac organogenesis in vitro: reestablishment of three-dimensional tissue architecture by dissociated neonatal rat ventricular cells. *Tissue Eng* **5**, 103–18.

42. Bursac N., Papadaki M., Cohen R. J., Schoen F. J., Eisenberg S. R., Carrier R., Vunjak-Novakovic G., and Freed L. E. (1999) Cardiac muscle tissue engineering: toward an in vitro model for electrophysiological studies. *Am J Physiol* **277**, H433–44.

43. Carrier R. L., Papadaki M., Rupnick M., Schoen F. J, Bursac N., Langer R., Freed L. E., and Vunjak-Novakovic G. (1999) Cardiac tissue engineering: cell seeding, cultivation parameters, and tissue construct characterization. *Biotechnol Bioeng* **64**, 580–9.

44. Li R. K., Jia Z. Q., Weisel R. D., Mickle D. A., Choi A., and Yau T. M. (1999) Survival and function of bioengineered cardiac grafts. *Circulation* **100**, II63–9.

45. Fink C., Ergun S., Kralisch D, Remmers U., Weil J., and Eschenhagen T. (2000) Chronic stretch of engineered heart tissue induces hypertrophy and functional improvement. *FASEB J* **14**, 669–79.

46. Leor J., Aboulafia-Etzion S., Dar A., Shapiro L., Barbash I. M., Battler A., Granot Y., and Cohen S. (2000) Bioengineered cardiac grafts: A new approach to repair the infarcted myocardium? *Circulation* **102**, 56–61.

47. Li R. K., Yau T. M., Weisel R. D., Mickle D. A., Sakai T., Choi A., and Jia Z. Q. (2000) Construction of a bioengineered cardiac graft. *J Thorac Cardiovasc Surg* **119**, 368–75.

48. Zimmermann W. H., Fink C., Kralisch D., Remmers U., Weil J., and Eschenhagen T. (2000) Three-dimensional engineered heart tissue from neonatal rat cardiac myocytes. *Biotechnol Bioeng* **68**, 106–14.

49. Eschenhagen T., Didie M., Heubach J., Ravens U., and Zimmermann W. H. (2002) Cardiac tissue engineering. *Transpl Immunol* **9**, 315–21.

50. Kofidis T., Akhyari P., Boublik J., Theodorou P., Martin U., Ruhparwar A., Fischer S., Eschenhagen T., Kubis H. P., Kraft T., Leyh R., and Haverich A. (2002) In vitro engineering of heart muscle: artificial myocardial tissue. *J Thorac Cardiovasc Surg* **124**, 63–9.

51. Krupnick A. S., Kreisel D., Engels F. H., Szeto W. Y., Plappert T., Popma S. H., Flake A. W., and Rosengard B. R. (2002) A novel small animal model of left ventricular tissue engineering. *J Heart Lung Transplant* **21**, 233–43.

52. Shimizu T., Yamato M., Isoi Y., Akutsu T., Setomaru T., Abe K., Kikuchi A., Umezu M., and Okano T. (2002) Fabrication of pulsatile cardiac tissue grafts using a novel 3-dimensional cell sheet manipulation technique and temperature-responsive cell culture surfaces. *Circ Res* **90**, e40.

53. Zimmermann W. H., Schneiderbanger K., Schubert P., Didie M., Munzel F., Heubach J. F., Kostin S., Neuhuber W. L., and Eschenhagen T. (2002) Tissue engineering of a differentiated cardiac muscle construct. *Circ Res* **90**, 223–30.

54. Eschenhagen T., Didie M., Munzel F., Schubert P., Schneiderbanger K., and Zimmermann W. H. (2002) 3D engineered heart tissue for replacement therapy *Basic Res Cardiol* **97**, I146–52.

55. Zimmermann W. H., Didie M., Wasmeier G. H., Nixdorff U., Hess A., Melnychenko I., Boy O., Neuhuber W. L., Weyand M., and Eschenhagen T. (2002) Cardiac grafting of engineered heart tissue in syngenic rats *Circulation* **106**, I151–7.

56. Kellar R. S., Landeen L. K., Shepherd B. R., Naughton G. K., Ratcliffe A., and Williams S. K. (2001) Scaffold-based three-dimensional human fibroblast culture provides a structural matrix that supports angiogenesis in infarcted heart tissue. *Circulation* **104**, 2063–8.

57. Vacanti J. P., Langer R., Upton J., and Marler J. J. (1998) Transplantation of cells in matrices for tissue regeneration. *Adv Drug Deliv Rev* **33**, 165–82.

58. Zimmermann W. H., and Eschenhagen T. (2003) Cardiac tissue engineering for replacement therapy. *Heart Fail Rev* **8**, 259–69.

59. Eschenhagen T., Fink C., Remmers U., Scholz H., Wattchow J., Weil J., Zimmermann W., Dohmen H. H., Schafer H., Bishopric N., Wakatsuki T., and Elson E. L. (1997) Three-dimensional reconstitution of embryonic cardiomyocytes in a collagen matrix: a new heart muscle model system. *FASEB J* **11**, 683–94.

60. Christman K. L., and Lee R. J. (2006) Biomaterials for the treatment of myocardial infarction. *J Am Coll Cardiol* **48**, 907–13.

61. Zammaretti P., and Jaconi M. (2004) Cardiac tissue engineering: regeneration of the wounded heart. *Curr Opin Biotechnol* **15**, 430–4.

62. Leri A., Kajstura J., Anversa P., and Frishman W. H. (2008) Myocardial regeneration and stem cell repair. *Curr Probl Cardiol* **33**, 91–153.

Part I

Isolation and Characterization of Cells for Myocardial Regeneration

Chapter 2

Bone Marrow-Derived Endothelial Progenitor Cells: Isolation and Characterization for Myocardial Repair

Masaaki Ii

Abstract

Vascular regeneration with bone marrow (BM) stem/progenitor cells is one of the promising therapeutic strategies for myocardial repair in cardiovascular diseases. Endothelial progenitor cells (EPCs) have demonstrated the beneficial effects on ischemia-induced myocardial damage promoting angiogenesis in ischemic tissue. Isolation of EPCs from BM or peripheral blood is required for the cell-based therapeutic approach. Among a variety of EPC isolation methods, specifically in the case of experimental animal models, we have shown our recently developed mouse-cultured EPC isolation protocol, which is viable for obtaining enough number of viable cells for not only in vitro but also in vivo experiments, and the protocol for human CD34+ cell (EPC rich cell population) isolation from peripheral blood and characterization by fluorescence-activated sorting (FACS) system.

Key words: Bone marrow, Stem/progenitor cell, Regeneration, Angiogenesis, Cell therapy, Cytokine, Growth factor, Ischemia, Myocardium

1. Introduction

Ischemic heart diseases are one of the major causes of death or hospitalization that are limiting our daily life. Therefore, a certain promising therapeutic approach for myocardial repair needs to be developed rather than conventional therapy such as medication alone. Revascularization is well known to be a crucial event during the recovery process in myocardium with ischemic injury. In the traditional view, revascularization of ischemic tissue was thought to occur through the migration and proliferation of mature endothelial cells in nearby tissues – a process called "angiogenesis"; however, recent evidences have indicated that the peripheral blood (PB) of adults contains bone marrow (BM)-derived cells

Randall J. Lee (ed.), *Stem Cells for Myocardial Regeneration: Methods and Protocols*, Methods in Molecular Biology, vol. 660, DOI 10.1007/978-1-60761-705-1_2, © Springer Science+Business Media, LLC 2010

with properties similar to those of embryonic angioblasts (1–4). These precursor cells have the potential to differentiate into mature endothelial cells and are collectively referred to as endothelial progenitor cells (EPCs). Emerging evidences indicate that the so-called EPC-participating vasculogenesis contributes significantly to postnatal neovascularization. These novel insights into molecular processes that contribute to the formation of blood vessels suggest a potential strategy for treatment of ischemic heart disease, namely, the transplantation of EPCs to induce neoangiogenesis in ischemic myocardium which also plays a critical role for myocardial repair including cardiac functional recovery.

The remainder of this chapter describes detailed protocols for EPC isolation from mouse BM and CD34+ cell (EPC rich cell population) isolation from human peripheral blood with the characterization not only as a distinct progenitor of endothelial cell (EC), which directly contributes to vasculogenesis (5,6), but also as a heterogeneous EC-like cell population which can promote angiogenesis indirectly (7,8).

2. Materials

2.1. Mouse BM Mononuclear Cell Isolation

1. 2,2,2-Tribromoethanol (Avertin[TR], Sigma, St. Louis, MO) is dissolved in tissue-culture water at 20 mg/mL with heat and stored in aliquots at 4°C protecting from light.
2. Chemical hair remover (Nair[TR], Church & Dwight Co., Inc., Lakewood, NJ).
3. 10% Povidone/Iodine prep pad (Dynarex PVP Iodine Prep Pads, Fisher Scientific, Pittsburgh, PA).
4. Scissors (Tungsten-Carbide Iris Scissors, Fine Science Tool: FST, Tokyo, Japan).
5. Forceps (Tissue Forceps Slim – 1 × 2 Teeth, FST).
6. Disposable 1 mL syringe with 23G needle (Terumo, Tokyo, Japan).
7. Disposable 3 mL syringe with 18G needle (Terumo).
8. 10 cm cell culture dishes (Corning, Lowell, MA).
9. Dissecting board (Neoprene Cork Board, Fisher Scientific).
10. Mortar (Agate Mortar and Pestle Sets 65 mm, Fisher Scientific).
11. 40 μm Cell Strainer (BD Falcon, Tokyo, Japan).
12. 5 mM EDTA/DPBS: Solution of 5 mM ethylenediamine tetraacetic acid (EDTA) in Dulbecco's Phosphate Buffered Saline (DPBS) (Invitrogen, Tokyo, Japan).

13. 50 mL and 15 mL conical tubes (BD Falcon).

14. Histopaque 1083^{TR} (Sigma) stored at 4°C protecting from light.

15. 0.8% (w/v) NH_4Cl with 0.1 mM EDTA: Ammonium chloride solution (StemCell Technologies, Vancouver, Canada).

2.2. Mouse BM Mononuclear Cell Culture

1. RepCell^{TR}: Temperature-responsive culture ware (CellSeed, Tokyo, Japan) (see Note 1).

2. Rat vitronectin (rVN, Sigma) is dissolved in 5 μg/mL (1 V/10 mL) of tissue-culture water and stored in 15 mL conical tube (BD Falcon) at 4°C.

3. Cell culture medium: EGM-2 medium (500 mL EBM-2^{TR} plus SingleQuots^{TR} of growth supplements without hydrocortisone, Lonza, Tokyo, Japan) stored at 4°C. Growth supplements (aqueous solution): 50 mL fetal bovine serum (FBS) (see Note 2), epidermal growth factor (EGF), vascular endothelial growth factor (VEGF), insulin-like growth factor-1 (IGF-1), basic fibroblast growth factor (bFGF), GA-1000 (Gentamicin Sulfate and Amphotericin-B).

2.3. Mouse EPC Characterization

2.3.1. EPC Culture Assay

1. Dulbecco's Phosphate Buffered Saline with calcium and magnesium (DPBS w/Ca&Mg, Invitrogen, Tokyo, Japan).

2. Acetylated Low Density Lipoprotein, labeled with 1,1′-dioctadecyl – 3,3,3′,3′-tetramethyl-indocarbocyanine perchlorate (200 μg/mL DiI-ac-LDL, Biological Technologies, Stoughton, MA) stored at 4°C.

3. Fluorescein-labeled GSL I isolectin B4 (ILB4) (1 μg/mL, Vector Laboratories, Burlingame, CA) stored at 4°C.

4. Two percent paraformaldehyde (PFA)/DBPS solution is made by diluting 4% PFA/DPBS (Wako, Osaka, Japan) with DPBS and stored at 4°C.

2.3.2. Fluorescent Immunocytochemistry

1. Microscope cover slips (24×40 mm) (Matsunami Glass, Osaka, Japan) and Lab-Tek 8-well chamber glass slides (Nalge Nunc, Naperville, IL).

2. Two percent PFA/DBPS solution is made by diluting 4% PFA/DPBS with DPBS (1:1) and stored at 4°C.

3. Antibody dilution buffer: 2% BSA in DPBS.

4. Quench solution: 50 mM NH_4Cl in DPBS.

5. Primary antibodies: rabbit antimouse endothelial nitric oxide synthase (eNOS) antibody (Sigma), rat antimouse Flk-1/vascular endothelial growth factor-receptor 2 (VEGF-R2) antibody (CHEMICON/Millipore, Billerica, MA), rabbit antimouse vascular endothelial cadherin (VE-cadherin) antibody

(Santa Cruz Biotechnology. Santa Cruz, CA), rabbit antivon Willebrand factor (vWF) antibody (CHEMICON), rat antimouse CD31 antibody (Santa Cruz Biotechnology), rat antimouse CD105 antibody (BD Bioscience, San Jose, CA), biotinylated goat anti-Tie2 antibody (R&D).

6. Secondary antibodies: Cy2 or Cy3 conjugated goat antirabbit IgG antibody, Cy2 or Cy3 conjugated goat antirat IgG antibody, Cy2 or Cy3 conjugated rabbit antigoat IgG antibody, FITC conjugated Streptoavidin (Jackson ImmunoResearch, West Grove, PA).

7. Permeabilization solution: 0.1% (v/v) Triton X-100 (Sigma) in DPBS.

8. Nuclear stain: 300 nM 4,6-diamidino-2-phenylindole (DAPI, Sigma) in DPBS.

9. Mounting medium: (Gel/Mount^{TR}, GeneTex, San Antonio, TX) stored at room temperature.

2.3.3. Tube Formation Assay

1. 24-well flat bottom clear cell culture plate (Corning).

2. Growth factor reduced (GFr) Matrigel^{TR} (BD Biosciences) stored at –20°C.

3. Cell culture medium: 500 mL EGM-2 medium (Lonza) (Subheading 2.2).

2.4. Human PB Mononuclear Cell Isolation

1. 50 mL and 15 mL conical tubes (BD Falcon).

2. 10 mL disposable syringe with 18G needle (Terumo).

3. Drip infusion kit with 18G i.v. catheter (Terumo).

4. Tourniquet (Velket^{TR} Tourniquets, VWR, West Chester, PA) and alcohol prep pad (VWR).

5. Solution of 5 mM EDTA in Ca and Mg-free DPBS (Invitrogen).

6. Histopaque 1077^{TR} (Sigma) stored at 4°C protecting from light.

7. NH_4Cl: Ammonium chloride solution (StemCell Technologies) stored at 4°C.

2.5. Human CD34+ Cell Isolation by MACS System

1. CD34 MultiSort Kit (# 130-056-701, Miltenyi Biotec, Tokyo, Japan) containing:

MACS MultiSort MicroBeads conjugated to monoclonal antihuman CD34 antibody. Isotype: mouse IgG1. Clone: QBEND/10.

MultiSort Release Reagent for enzymatic release of MultiSort MicroBeads bound to the cell surface.

MultiSort Stop Reagent to inhibit the release reaction for further separations.

2. FcR-Blocking Reagent (Miltenyi Biotec).

3. MiniMACS™ Separator with MS Column Adaptor and MS Column (Miltenyi Biotec).

4. 30 µm nylon mesh (Preseparation Filter # 130-041-407, Miltenyi Biotec).

5. DPBS supplemented with 0.5% (w/v) bovine serum albumin (BSA) and 2 mM EDTA by diluting MACS^{TR} BSA Stock Solution with Ca and Mg free DPBS at 1:20 ratio (Miltenyi Biotec).

6. 15 mL conical tube (BD Falcon).

2.6. Human CD34$^+$ Cell Characterization by FACS Analysis

1. FACS Calibur^{TR} 4A (BD Biosciences).

2. 1.5 mL micro tubes: (Eppendorf 3180, Fisher).

3. Buffer: Hanks Balanced Salt Solution (HBSS, Invitrogen) supplemented with 3% FBS.

4. PI buffer: HBSS supplemented with 2 µg/mL Propidium Iodine (PI Solution, Sigma) and 3% FBS.

5. FcR Blocking Reagent, human (Miltenyi Biotec).

6. PE-conjugated antimouse IgG$_{1K}$ (BD Pharmingen).

7. APC-conjugated antimouse IgG$_{1K}$ (BD Pharmingen).

8. FITC-conjugated antimouse IgG$_{1K}$ (BD Pharmingen).

9. PE-conjugated antihuman CD34 mouse IgG$_{1K}$ antibody (BD Pharmingen).

10. APC-conjugated antihuman CD133 mouse IgG$_{1K}$ antibody (Miltenyi Biotec).

11. APC-conjugated antihuman CD117/c-kit mouse IgG$_{1K}$ antibody (BD Pharmingen).

12. FITC-conjugated antihuman CD31 mouse IgG$_{1K}$ antibody (BD Pharmingen).

13. Purified antihuman VEGF-R2/KDR mouse IgG$_{1K}$ antibody (Sigma).

14. Biotin antimouse IgG$_{1K}$ (BD Biosciences).

15. APC-conjugated Streptoavidin (BD Pharmingen).

16. APC-conjugated antihuman Tie2 (DAKO, Tokyo, Japan).

17. Purified antihuman VE-cadherin mouse IgG$_{2aK}$ (HyCult Biotechnology, Uden, Netherlands).

18. Biotin antimouse IgG$_{2aK}$ (BD Biosciences).

19. FACS Calibration beads (Calibrite 3 beads, BD Calibrite).

20. FACS washing solutions: BD FACS Flow solution, BD FACS Clean solution, BD FACS Rinse solution (BD FACS).

3. Methods

Since the discovery of BM-derived circulating EPCs as a CD34[+] cell fraction in adult human peripheral mononuclear cells (1), we and others have been using CD34[+] or CD133[+] cells alone or the combination of KDR[+] as a EPC-rich cell fraction (9) in peripheral blood in clinical studies for chronic ischemic limb and heart diseases. On the other hand, a distinct definition of mouse EPCs by cell surface markers has not been determined, and investigators currently isolate mouse EPCs with their own criteria, i.e., any combination of stem/progenitor markers, sca-1, c-kit, CXCR4, or CD34 and endothelial markers, CD31, VE-cadherin, Tie2, or Flk-1, etc. in Lineage, CD11b, or CD45 negative cell population by FACS or MACS sorting method. (9) Moreover, one of the disadvantages in the freshly isolated mouse EPCs is the difficulty in obtaining enough number of cells for any experiment, specifically in vivo study. In contrast, although mouse-cultured EPCs isolated from BM are quite heterogeneous, enough number of EPCs can be obtained by this method and are practical for any experiment.

In this method, since the definition of mouse EPCs by cell surface markers is controversial, the author focuses on cultured EPCs for mouse cells. The author also focuses on CD34[+] cells for human EPCs that are currently used as a cell source of autologous BM-derived EPC transplantation therapy in patients with not only chronic hindlimb ischemia in our institute but also chronic ischemic cardiovascular diseases in other institutes. The author particularly introduces recently developed culturing technique for mouse EPC isolation by which most of the other differentiated cell types, i.e., monocyte/macrophage, mature endothelial cell and fibroblast can be eliminated comparing with an originally developed method (1,7,10), and a sensitive characterization method for human EPCs (CD34[+] cells) by FACS analysis.

3.1. Mouse BM Mononuclear Cell Isolation

1. After anesthetizing a mouse by intraperitoneal injection of 0.5 mL Avertin[TR] solution, take as much blood as possible with a 23G needle-1 mL disposable syringe from the heart percutaneously; remove all back hair by shaving and using Nair[TR] followed by putting povidone/iodine on the back.

2. Fix the mouse on dissecting board in abdominal position and remove all bones (all limbs, hip, sternum, and back bone) with the muscle using sterile forceps and scissors roughly and soak in 5-mM EDTA/DPBS solution with a 10-cm cell culture dish.

3. Remove muscle/spinal cord from all bones including spine and soak the cleaned bones in new 5-mM EDTA/DPBS solution with a 10-cm cell culture dish.

4. Transfer all bones to sterile mortar and mince them with sterile scissors into 3–4 mm sized pieces.

5. Grind the minced bones by stone stick in 5-mM EDTA/DPBS solution (see Note 3).

6. Collect supernatant (BM cell suspension) with a 18G needle-5-mL disposable syringe, and transfer to 40-µm cell strainer with a 50-mL conical tube.

7. Repeat step 6 with fresh 5-mM EDTA/DPBS solution until the supernatant becomes clear (see Note 4).

8. Divide the cell suspension into two 15-mL conical tubes with 4 mL of Histopaque1083 gently overlaying cell suspension on Hitopaque1083 (see Note 5).

9. Centrifuge samples at $1,100 \times g$ speed for 20 min at room temperature (RT) without brake.

10. Collect cloudy mononuclear cell layer with 18G needle-3 mL disposable syringe (see Note 6) and transfer to new15 mL conical tubes, and fill the tubes with 5-mM EDTA/DPBS solution.

11. Centrifuge samples at $1,150 \times g$ speed for 5 min at 4°C with low brake.

12. Discard the supernatant and add 1 mL of 5-mM EDTA/DPBS solution suspending cell pellet by pipetting.

13. Add 13 mL of 5-mM EDTA/DPBS solution and centrifuge at $240 \times g$ speed for 5 min at 4°C with low brake.

14. Discard (platelet containing) supernatant and add 1 mL of 5 mM EDTA/DPBS solution suspending pellet by pipetting.

15. Add 4 mL of NH_4Cl mixing well and incubate for 10 min at 4°C (on ice).

16. Add 9 mL of 5 mM EDTA/DPBS solution and centrifuge at $350 \times g$ speed for 15 min at 4°C with low brake.

17. Discard supernatant and add 2 mL of EGM-2 medium in each tube.

18. Combine two samples into one 50 mL conical tube (total 4 mL), and count the cell number.

3.2. Mouse BM Mononuclear Cell Culture

1. Add 3 mL of 5 µg/mL rVN solution in RepCell dish covering all surface and quickly return the solution to original 15 mL conical tube, and incubate the rVN-coated dish at RT in clean bench until the dish dries up.

2. Add appropriate amount of DMEM supplemented with 10% FBS in mouse BM mononuclear cell suspension (4 mL in 50 mL conical tube, Subheading 3.1) and seed the cells dividing into rVN-coated RepCell dishes as indicated in Table 1.

3. After 24 hours in culture in CO_2 incubator (5% CO_2 at 37°C), transfer all floating cells with whole culture medium to new 50 mL tube, and leave the RepCell dish at RT for 15–20 min (see Note 7).

4. After centrifugation of the 50 mL tube (step 3) at $300 \times g$ speed for 5 min, the pelleted cells are suspended with EGM-2 medium.

5. Remove the attached cells by washing/pipetting with DPBS three times, and put floating cell suspension (step 4) into original RepCell dish.

6. After 3 days in culture, change the medium with fresh EGM-2 medium removing floating cells and keep culturing in CO_2 incubator for further 3 days.

7. On day 7, after replacing medium with DPBS, incubate the RepCell dishes with attached cells at RT for 15–20 min followed by washing/pipetting with the left DPBS, and collect all detached cell solution in 15 mL conical tubes.

8. After centrifugation at $300 \times g$ speed for 5 min, pelleted cells can be used for any experiment as EPC-rich cell population.

3.3. Mouse EPC Characterization

3.3.1. EPC Culture Assay

1. Seed mouse cultured EPCs obtained from BM mononuclear cells according to the above protocol (Subheading 3.2) onto 8-well chamber glass slide at a density of 2×10^4/well with 250 µL of EGM-2 medium, and further culture until the cells attach on the bottom (approximately 4–6 h).

2. Add 4 µL of DiI-ac-LDL solution into each well and shake the slide gently and incubate for 4 h.

3. Wash the samples with DPBS w/Ca&Mg three times and fix the cells with 100 µL of 2% PFA/DPBS solution for 15 min at RT.

4. Remove 2% PFA/DPBS solution and wash the samples with DPBS w/Ca&Mg three times, and then incubate the fixed cells with 100 µL of FITC-ILB4 solution for 1 h at 37°C.

Table 1
Optimized cell number in RepCell^{TR} cell culture dish for EPC culture

Dish size	Cell number	Medium (mL)
10 cm diameter	$60–80 \times 10^6$	16–20
6 cm diameter	$20–30 \times 10^6$	4–6

5. Remove FITC-ILB4 solution and wash the samples with DPBS w/Ca&Mg three times, and then incubate with 100 µL of DAPI solution for 10 min at RT for nuclear staining.

6. Rinse samples with DPBS w/Ca&Mg and remove gaskets carefully from the glass slides using a razor blade.

7. Mounting medium and a cover-slip are added to the glass slide followed by sealing the sample with nail varnish. The sample slides can be viewed immediately after the varnish is dry, or be stored in the dark at 4°C for up to a month.

8. Slides are viewed under fluorescent microscopy. Excitation at 549 nm includes the DiI fluorescence (red emission) for ac-LDL uptake, excitation at 490 nm includes the FITC fluorescence (green emission), and excitation at 350 nm includes the DAPI fluorescence (blue emission). Software (Adobe Photoshop[TR] CS3, Adobe Systems, San Jose, CA) can be used to overlay fluorescent images to detect double positivity of DiI and FITC signals, which indicates one of the EPC characteristics. An example result is shown in Fig. 1a.

3.3.2. Fluorescent Immunocytochemistry

1. Seed mouse cultured EPCs obtained from BM mononuclear cells according to the above protocol (Subheading 3.2) onto 8-well chamber glass slide at a density of 2×10^4/well with 200 µL of EGM-2 medium, and further culture until the cells attach on the bottom (It takes approximately 4–6 h).

2. Wash samples with DPBS three times and replace DPBS with 100 µL of 2% PFA/DPBS solution for 15 min at RT to fix the cells.

3. Remove 2% PFA/DPBS solution and wash the samples with DPBS three times.

4. Residual PFA is quenched by incubation in 50 mM NH_4Cl solution for 10 min at RT, followed by further washing with DPBS three times.

5. Cells are permeabilized by incubation in Triton X-100 solution for 5 min at RT (see Note 8), and then rinsed with DPBS three times.

6. Samples are blocked by incubation in antibody dilution buffer for 1 h at RT, and the plastic chambers should be removed leaving gasket on the slide.

7. Remove blocking solution and replace with primary antibodies: anti-eNOS antibody (1:200), anti-vWF antibody (1:500), anti-VE-cadherin(1:50), anti-Tie2 antibody (1:50), anti-Flk-1 antibody (1:500), anti-CD31 antibody (1:50), and anti-CD105 antibody (1:20) in antibody dilution buffer at 4°C over night.

8. Remove primary antibodies and wash samples three times for 5 min each with DPBS.

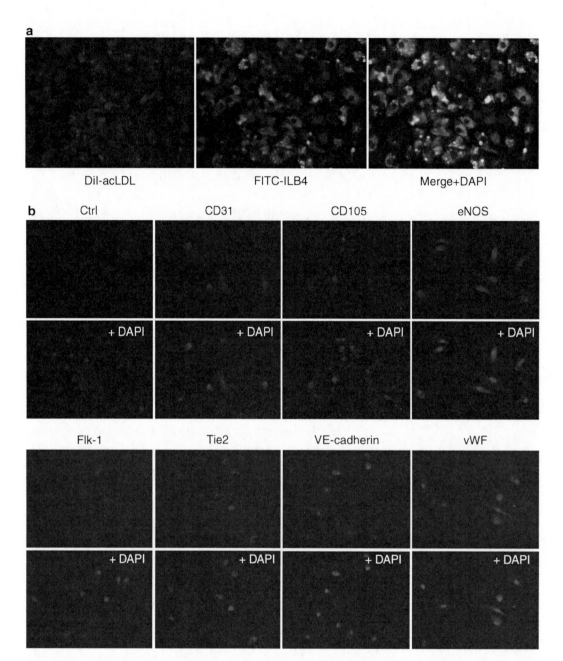

Fig. 1. EPC characteristics by EPC culture assay and immunocytochemistry with mouse BM mononuclear cells. (a) Endothelial characteristics are assessed by EPC culture assay. Dil-acLDL uptake cells are observed in red and FITC-ILB4 binding cells are observed in green under fluorescent microscope. Double positive (*yellow in merged images*) cells are considered as BM-derived endothelial lineage cells, namely, EPCs. (b) Immunocytostaining with a series of endothelial markers confirms endothelial phenotype of EPCs. CD31, CD105, eNOS, Flk-1, Tie2, VE-cadherin, and vWF are stained in *red* with DAPI (*blue*). Over 80% of total adherent cells are positive for each marker except for Flk-1 (40%) compared with IgG staining control (Ctrl). Original magnification: ×400.

9. Secondary antibodies are prepared at 1:500 in antibody dilution buffer and added to the samples for 30 min at RT.

10. Remove secondary antibodies followed by DPBS washing for three times and DAPI solution (1:5000) is added to for 10 min at RT for nuclear staining.

11. Rinse samples with DPBS and remove gaskets carefully from glass slides using a razor blade.

12. Mounting medium and a cover-slip are added to the glass slide followed by sealing samples with nail varnish. The sample slides can be viewed immediately after the varnish is dry, or be stored in the dark at 4°C for up to a month.

13. Slides are viewed under fluorescent microscopy. Excitation at 543 nm includes the Cy3 fluorescence (red emission), excitation at 490 nm/488 nm includes the FITC/Cy2 fluorescence (green emission), respectively, and excitation at 350 nm includes the DAPI fluorescence (blue emission). Software (Adobe Photoshop^{TR} CS3, Adobe Systems) can be used to adjust the fluorescent signal intensity in each image to detect immunoreactivity of endothelial markers. An example result is shown in Fig. 1b.

3.3.3. Colony/Tube Formation Assay

1. GFrMatrigel^{TR} is prepared on ice or in refrigerator (4°C) for 3 h before assay.

2. Coat 24-well culture dish with 200 µl of GFrMatrigel^{TR}/each well on ice by cold pipette tip.

3. Incubate the 24-well culture dish at 37°C for 30 min.

4. Seed mouse cultured EPCs obtained from BM mononuclear cells according to the above protocol (Subheading 3.2) at a density of 5×10^4/well with 500 µl of EGM-2 medium onto growth factor-reduced GFrMatrigel^{TR}-coated dish.

5. Samples are cultured in CO_2 incubator (5% CO_2 at 37°C) for 48 h (see Note 9), and the cells in colony and tube formation are viewed under phase contrast microscope.

6. After capturing images by a computer-assisted software with microscope, the images are analyzed using a Image J^{TR} software for measuring total tube length per colony and averaged. Tube length accounts for vasculogenic capacity of the cells that have endothelial-like cell phenotype. An example result is shown in Fig. 2.

3.4. Human Peripheral Blood Mononuclear Cell Isolation

1. Six 50 mL conical tubes with each 15 mL of 5 mM EDTA/DPBS solution are prepared for blood collection.

2. After wearing a tourniquet on either right or left upper arm, take blood using a drip infusion kit with 18G i.v. catheter, and add 35 mL of blood in each 50 mL conical tube (total 210 mL).

Colony Formation   Tube Formation

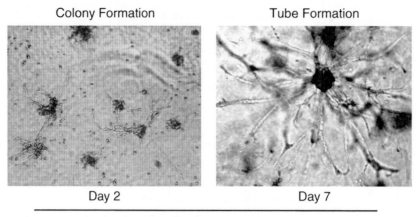

Day 2   Day 7

Time after Assay

Fig. 2. Colony and Tube formation assay with Day 7 mouse BM-derived cultured EPCs. Cultured EPCs are obtained from BM mononuclear cells after 7 days in culture in EPC differentiation medium (EGM-2) and further cultured on GFrMatrigel[TR] for 48 h (Day 2) and 7 days (Day 7). After forming colonies with spicula, remarkable elongated tube-like structure originated from colonies is observed, indicating vascular forming capacity. On Day 2, after culture with 10%FBS/DMEM medium, attached cells form colonies, but not tube-like structure. (data not shown).

3. Prepare eight 50 mL conical tubes with each 12.5 mL Histopaque 1077[TR] at RT.

4. Divide the blood diluted with 5 mM EDTA/DPBS solution (step 3) into eight 50 mL conical tubes with 12.5 mL of Histopaque1077[TR] (step 4) gently overlaying cell solution on Hitopaque1077 (see Note 5).

5. Centrifuge the samples at $1,100 \times g$ speed for 30 min at RT without brake.

6. Collect cloudy mononuclear cell layer with 18G needle-10 mL disposable syringe (see Note 6) and transfer to four new 15 mL conical tubes, and fill the tubes with 5 mM EDTA/DPBS solution.

7. Four 50 mL conical tubes with 10 mL of 5 mM EDTA/DPBS solution are prepared, and each mononuclear cell sample (step 7) is added to the four 50 mL conical tubes.

8. Centrifuge samples at $1,150 \times g$ speed for 5 min at 4°C with low brake.

9. Discard supernatant and add 10 mL of 5 mM EDTA/DPBS solution suspending cell pellet by pipetting.

10. Add 20 mL of 5 mM EDTA/DPBS solution and centrifuge at $240 \times g$ speed for 5 min at 4°C with low brake.

11. Discard supernatant and add 2 mL of 5 mM EDTA/DPBS solution suspending cell pellet by pipetting.

12. Add 8 mL of NH_4Cl mixing well and incubate at 4°C (on ice) for 10 min.

13. Add 40 mL of 5 mM EDTA/DPBS solution and centrifuge at $350 \times g$ speed for 15 min at 4°C with low brake.

14. Discard supernatant and add 2 mL of EGM-2 medium in each tube.

15. Combine four samples into one 15 mL conical tube (total 8 mL), and count cell number.

3.5. Human CD34+ Cell Isolation

1. Resuspend 2×10^8 human PB mononuclear cells obtained according to the above protocol (Subheading 3.4) with buffer in a final volume of 600 µL.

2. Add 200 µL of FcR Blocking Reagent to the cell suspension to inhibit unspecific or Fc-receptor binding of CD34 MultiSort MicroBeads to nontarget cells.

3. Label cells by adding 200 µL of CD34 MultiSort MicroBeads, mix well, and incubate for 30 min at 4°C (see Note 10).

4. Wash cells by adding 1 mL of buffer, centrifuge at $300 \times g$ for 10 min, remove supernatant completely, and then resuspend in 500 µL of buffer.

5. Place an MS Column (combined with the appropriate Column Adapter) in the magnetic field of the MACS Separator.

6. Prepare the column by washing with 500 µL of buffer.

7. Apply cell suspension on top of the column and let the unlabeled cells pass through followed by rinsing with 500 µL of buffer three times.

8. Remove column from separator and place column on a 15 mL conical tube.

9. Add 1 mL of buffer onto the column and flush out magnetically labeled fraction using the plunger supplied with the column (see Note 11).

10. To achieve a higher purity, apply magnetically labeled fraction onto a new freshly prepared column, and let the unlabeled cells pass through followed by rinsing with 500 µL of buffer three times.

11. The magnetically labeled cell fraction are eluted and collected in the conical tube.

12. Incubate magnetically labeled cells with 20 µL of MACS^{TR} MultiSort Release Reagent/mL of cell suspension for 10 min at 4°C.

13. Separate cells over a new MS Column to remove any remaining magnetically labeled cells, and prepare the column by washing with 500 µL of buffer.

14. Apply cell suspension on top of the column and let the unlabeled cells pass through followed by rinsing with 500 µL of buffer three times.

15. Wash cells from released fraction removing supernatant completely, and resuspend the cell pellet in buffer in a final volume of 40 µL.

16. Add 60 µL of MACS MultiSort Stop Reagent/40 µL of cell suspension and mix well.

17. The cells in 100 µL of buffer are magnetically sorted CD34+ cells.

3.6. Human CD34+ Cell Characterization by FACS Analysis

1. Freshly isolated CD34+ cells (Subheading 3.4, step 17) are centrifuged at $300 \times g$ for 10 min. After removing supernatant, suspend the cell pellet with buffer (3% FBS/HBSS) at a concentration of 8×10^6 cells/mL, and make ten samples in 1.5 mL micro tubes with 2×10^5 cells/25 µL of cell solution each.

2. Add 2 µL of FcR Blocking Reagent in each cell sample and mix well, and then incubate for 30 min at 4°C.

3. Add antibodies in each cell solution as indicated in Table 2 and incubate for 30 min at 4°C followed by addition of 1 mL buffer and mixing (see Note 12).

4. After centrifugation at $2,000 \times g$ speed for 2 min at 4°C, wash the cells with 1 mL buffer + by pipetting and repeat this step twice.

5. Remove supernatant and suspend the cells with 300 µL of PI buffer each.

6. The samples are ready for FACS analysis with FACS Calibur 4A.

7. For calibration, after turning on the power and starting FACS Comp software, uptake FACS calibration beads (Calibrite 3 beads) in the machine by pressing [RUN] button.

8. After the automatic calibration, wash the circuit with FACS Flow solution in the same manner as step 7 (see Note 13).

9. After starting CellQuestPro^TR software (BD Biosciences), uptake the sample #1 (Table 2) in the machine by pressing [RUN] button and the sample is analyzed automatically.

10. For compensation, adjust voltage levels for the next PE, APC, and FITC measurement and also for PI on the dot plot screen in the result of sample tube #1 (Table 2) (negative controls).

11. Uptake samples #2, #3, and #4 (Table 2) in the machine by pressing [RUN] button and the samples are analyzed automatically.

12. In the results of sample tubes #2 (FITC control), #3 (PE control), and #4 (APC control), adjust compensation values at FL3-FL1, FL3-FL2, and FL3-FL4 channels to avoid interference of FITC, PE, and APC signals to PI signal, respectively (see Note 14).

Table 2
Combination of antibodies with human CD34+ Cells for FACS analysis

Tube #	IgG or antibodies (antibody 1 + antibody 2)	Volume
1	FITC-mouse IgG$_{1K}$ (isotype control for CD31,CD34) + PE-mouse IgG$_{1K}$ (isotype control for CD34, KDR) + APC-mouse IgG$_{1K}$ (isotype control for CD133, CD117, Tie2)	2 μL + 2 μL + 2 μL
2	FITC antihuman CD45 (for FITC compensation)	2 μL
3	PE-antihuman CD45 (for PE compensation)	2 μL
4	APC-antihuman CD45 (for APC compensation)	2 μL
5	PE-antihuman CD34 + APC-antihuman CD133	2 μL + 2 μL
6	PE-antihuman CD34 + APC-antihuman CD117/c-kit	2 μL + 2 μL
7	FITC-antihuman CD31 + APC-mouse antihuman Tie2	2 μL + 2 μL
8	APC-Streptoavidin + Biotin-mouse IgG$_{2aK}$ (isotype control for VE-cadherin) + FITC-mouse IgG$_{1K}$	2 μL + 2 μL + 2 μL
9	FITC-antihuman CD31 + APC-Streptoavidin + Biotin-mouse antihuman VE-cadherin	2 μL + 2 μL + 2 μL
10	APC-Streptoavidin + Biotin-antimouse IgG$_{1K}$ + Purified mouse IgG1$_{1K}$ (for isotype control for KDR) + PE-mouse IgG$_{1K}$	2 μL + 2 μL + 2 μL + 2 μL
11	PE-antihuman CD34 + APC-Streptoavidin + Biotin-antimouse IgG$_{1K}$ + Purified mouse antihuman VEGF-R2/KDR	2 μL + 2 μL + 2 μL + 2 μL

13. For sample analysis, input the cell number (10,000) to be counted at "Acquisition & Storage" area on the screen.

14. Uptake samples #5, #6, and #7 (Table 2) in the machine by pressing [RUN] button and save the automatically analyzed data.

15. Next, uptake the sample #8 (Table 2) in the machine by pressing [RUN] button and the sample is analyzed automatically.

16. Adjust voltage levels for the next APC and FITC measurement and also for PI on the dot plot screen in the result of sample tube #8 (Table 2) (negative controls).

17. For sample analysis, input the cell number (10,000) to be counted at "Acquisition & Storage" area on the screen.

18. Uptake samples #9 (Table 2) in the machine by pressing [RUN] button and save the automatically analyzed data.

19. Adjust voltage levels for the next APC and FITC measurement and also PI on the dot plot screen in the result of sample tube #10 (Table 2) (negative controls).

20. Uptake samples #11 (Table 2) in the machine by pressing [RUN] button and save the automatically analyzed data.

21. After all measurements, circuit in the machine should be cleaned with FACS Clean solution and rinsed with FACS Rinse solution and be shut down.

22. For data analysis, select "Analysis" for dot plot on the CellQuest Pro software screen.

23. After retrieving the saved data, set up a gate in the targeted cell fraction and develop the gated part in another dot plot graph.

24. In the dot plot graph of negative control samples, set quadrant location on PE/APC or FITC/APC axis until the percent of PE/APC or FITC/APC negative control shows 100% negative for PE, FITC, and APC signals.

25. Apply the same values of quadrant location as negative control to the dot plot graph of the samples (#1, #2, #3, #4, #8 and #10), and then the percent of targeted cell fractions in total CD34+ cells are automatically calculated. An example result is shown in Fig. 3.

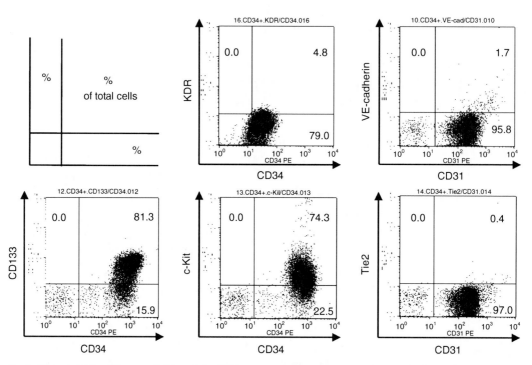

Fig. 3. Characterization of human CD34+ cells by FACS analysis. CD34+ cells are isolated by MACS system and the phenotype is assessed by FACS with stem/progenitor and endothelial markers. Although little cell population was positive for Flk1, VE-cadherin, and Tie2 that are generally expressed in well-differentiated ECs, high expressions of CD31, c-kit, and CD133 were observed in CD34+ cells, suggesting that most of the freshly isolated CD34+ cells were immature endothelial progenitors.

4. Notes

1. RepCell^{TR} (or UpCell^{TR}) (11) allow us to detach any strong adherent cells from culture dish easily without trypsinization. Since trypsinization more or less affects (reduces) cell surface marker expressions and the viability, this temperature-sensitive cell culture dish (RepCell^{TR} or UpCell^{TR}) is suitable for harvesting cells specifically for FACS analysis.

2. Since FBS strongly affects EPC differentiation to endothelial lineage depending on the lot and concentration in culture medium, the same lot of FBS should be used in all subsequent EPC-related assays in vitro.

3. To avoid BM cell death by grinding minced bone pieces, the procedure of gentle frequent tapping and rotation by stone stick is important for preserving cell viability. If white pellet can be observed with red blood cell pellet after the first centrifugation (step 9), dead mononuclear cells are in the white pellet, and a reduced number of mononuclear cells with less viability will be collected. Pellet color should be red at step 9.

4. Twenty milliliters of final cell suspension makes two samples of 15 mL tubes and is convenient for centrifugation (step 8).

5. For successful isolation of mononuclear cells, the temperature of Histopaque 1083^{TR} should be similar to RT, so Histopaque 1083^{TR} should be prewarmed at RT before step 8.

6. Collect all cell pellets on the wall of conical tube as well as mononuclear cells. The pellet also contains a number of mononuclear cells.

7. If cell detachment is not enough after 10–15 min incubation at RT, RepCell dish can be incubated in refrigerator for 5–10 min alternatively.

8. Triton-X100 treatment is only needed for staining of cytoplasmic molecules (eNOS and vWF). For cell surface molecule (CD31, Tie2, Flk-1, VE-cadherin, CD-105) staining, this treatment is not necessary.

9. Colony and tube formation can be observed 24 and 48 h after incubation, respectively. These findings can also be observed up to 14 days after incubation. Appropriate time points for observation should be determined by investigators.

10. Work fast, keep cells cold, and use precooled solutions. This procedure will prevent capping of antibodies on the cell surface and nonspecific cell labeling.

11. For optimal performance, it is important to obtain a single-cell suspension. Pass cells through 30 μm nylon mesh to remove cell clumps which may clog the column.

12. For staining in biotin-avidine system, samples should be incubated with each antibody separately, i.e., incubate the sample with biotin-antimouse IgG followed by washing step (step 4) and then do with APC-streptoavidine.

13. This washing step for circuit should be performed in each sample measurement.

14. Channel for each fluorescent is set up as: FL1 = FITC, FL2 = PE, FL3 = PI, FL4 = APC. To detect dead cells (PI positive cells) and avoid overlap of PI and other fluorescent, i.e., FITC, PE, or APC for measurement, adjustment of compensation value for PI signal should be performed at this step again.

Acknowledgments

The author would like to thank Miki Horii and Ayumi Yokoyama for technical assistance.

References

1. Asahara T, Murohara T, Sullivan A, Silver M, van der Zee R, Li T, Witzenbichler B, Schatteman G, Isner JM. Isolation of putative progenitor endothelial cells for angiogenesis. *Science*. 1997;275(5302):964–967.

2. Gehling UM, Ergun S, Schumacher U, Wagener C, Pantel K, Otte M, Schuch G, Schafhausen P, Mende T, Kilic N, Kluge K, Schafer B, Hossfeld DK, Fiedler W. In vitro differentiation of endothelial cells from AC133-positive progenitor cells. *Blood*. 2000; 95(10):3106–3112.

3. Peichev M, Naiyer AJ, Pereira D, Zhu Z, Lane WJ, Williams M, Oz MC, Hicklin DJ, Witte L, Moore MA, Rafii S. Expression of VEGFR-2 and AC133 by circulating human CD34(+) cells identifies a population of functional endothelial precursors. *Blood*. 2000;95(3): 952–958.

4. Quirici N, Soligo D, Caneva L, Servida F, Bossolasco P, Deliliers GL. Differentiation and expansion of endothelial cells from human bone marrow CD133(+) cells. *Br J Haematol*. 2001;115(1):186–194.

5. Kawamoto A, Iwasaki H, Kusano K, Murayama T, Oyamada A, Silver M, Hulbert C, Gavin M, Hanley A, Ma H, Kearney M, Zak V, Asahara T, Losordo DW. CD34-positive cells exhibit increased potency and safety for therapeutic neovascularization after myocardial infarction compared with total mononuclear cells. *Circulation*. 2006;114(20): 2163–2169.

6. Losordo DW, Schatz RA, White CJ, Udelson JE, Veereshwarayya V, Durgin M, Poh KK, Weinstein R, Kearney M, Chaudhry M, Burg A, Eaton L, Heyd L, Thorne T, Shturman L, Hoffmeister P, Story K, Zak V, Dowling D, Traverse JH, Olson RE, Flanagan J, Sodano D, Murayama T, Kawamoto A, Kusano KF, Wollins J, Welt F, Shah P, Soukas P, Asahara T, Henry TD. Intramyocardial transplantation of autologous CD34+ stem cells for intractable angina: a phase I/IIa double-blind, randomized controlled trial. *Circulation*. 2007;115(25): 3165–3172.

7. Ii M, Nishimura H, Iwakura A, Wecker A, Eaton E, Asahara T, Losordo DW. Endothelial progenitor cells are rapidly recruited to myocardium and mediate protective effect of ischemic preconditioning via "imported" nitric oxide synthase activity. *Circulation*. 2005;111(9): 1114–1120.

8. Urbich C, Aicher A, Heeschen C, Dernbach E, Hofmann WK, Zeiher AM, Dimmeler S. Soluble factors released by endothelial progenitor cells promote migration of endothelial cells and cardiac resident progenitor cells. *J Mol Cell Cardiol*. 2005;39(5): 733–742.

9. Jujo K, Ii M, Losordo DW. Endothelial progenitor cells in neovascularization of infarcted myocardium. *J Mol Cell Cardiol.* 2008; 45(4):530–544.

10. Ii M, Takenaka H, Asai J, Ibusuki K, Mizukami Y, Maruyama K, Yoon YS, Wecker A, Luedemann C, Eaton E, Silver M, Thorne T, Losordo DW. Endothelial progenitor thrombospondin-1 mediates diabetes-induced delay in reendothelialization following arterial injury. *Circ Res.* 2006;98(5):697–704.

11. Yanase Y, Suzuki H, Tsutsui T, Uechi I, Hiragun T, Mihara S, Hide M. Living cell positioning on the surface of gold film for SPR analysis. *Biosens Bioelectron.* 2007;23(4): 562–567.

Chapter 3

Umbilical Cord Blood Stem Cells for Myocardial Repair and Regeneration

Nicholas Greco and Mary J. Laughlin

Abstract

Cardiovascular disease remains a major cause of morbidity and mortality with substantial economic cost. There remains a need for therapeutic improvement for patients *refractory* to revascularization and those who redevelop occlusions following revascularization. Early evidence linked age-associated reductions in the levels of circulating marrow-derived hematopoietic stem cells (HSC), characterized by expression of early HSC markers CD133 and CD34, with the occurrence of cardiovascular events and associated death. Heart tissue has the endogenous ability to regenerate through the activation of resident cardiac stem cells or through recruitment of a stem cell population from other tissues, such as bone marrow. A number of clinical trials have utilized patient-derived autologous bone marrow-derived cells or whole BM uncultured mononuclear cells (MNC) infused or injected locally to augment angiogenesis. In most cases of treating animal models with human cells, the frequency of stem cell engraftment, the subsequent number of newly generated cardiomyocytes and vascular cells, and the augmentation of endogenous microvascular collateralization, either by deposition, transdifferentiation, and/or by cell fusion, appear to be too low to explain the significant cardiac improvement. Initially, it was hypothesized that cell therapy may work by cell replacement mechanisms, but recent evidence suggests alternatively that cell therapy works by providing trophic support to the injured tissues. An alternative hypothesis is that the transplanted stem cells release soluble cytokines and growth factors (i.e., paracrine factors) that function in a paracrine fashion, contributing to cardiac repair and regeneration by inducing cytoprotection and neovascularization. Another hypothesis which may also be operative is that cell therapy may mediate endogenous regeneration by the activation of resident cardiac stem cell. Well-established clinical trials have used cord blood for the treatment of hematological malignances (e.g., leukemia, lymphoma, myeloma) and nonmalignancies (e.g., in born errors of metabolism, sickle cells anemia, autoimmune diseases), but further advances in other areas of regenerative medicine (e.g., cardiac repair) will directly benefit with the use of cord blood. These clinical outcomes demonstrate that effector cells may be delivered by an allogeneic approach, where strict tissue matching may not be necessary and treatment may be achieved by making use of the trophic support capability of cell therapy and not by a cell replacement mechanism.

Key words: Revascularization, Bone marrow, Umbilical cord blood, Regenerative, Paracrine

Randall J. Lee (ed.), *Stem Cells for Myocardial Regeneration: Methods and Protocols*, Methods in Molecular Biology, vol. 660, DOI 10.1007/978-1-60761-705-1_3, © Springer Science+Business Media, LLC 2010

1. Introduction

The concept of regenerative medicine is relatively new, but both bone marrow (BM) and UCB stem cells have been used extensively in clinical procedures involving hematopoietic and immune system reconstitution or transplantation for cancer patients (i.e., leukemia, lymphoma, myeloma) whose bone marrow has been ablated by chemotherapy and radiotherapy. This emerging clinical application of the 1950s became a successful state-of-procedure in the late 1970s with a better clinical and scientific understanding of human histocompatibility matching and a further understanding that cells transplanted from a human volunteer had to be "matched" immunologically to the human recipient. Cardiovascular disease is the leading cause of morbidity and mortality in the USA accounting for approximately one million deaths per year in spite of therapeutic intervention advances (1). Occlusion of coronary arteries accounts for the majority of these deaths. A significant number of these patients with areas of viable myocardium in risk by impaired perfusion are poor targets for conventional revascularization techniques (2) Clinical studies with stem cells are concerned preferably with three clinically relevant situations: therapy for acute myocardial infarction, therapy for chronic myocardial infarction, and therapy for congestive heart failure (dilated cardiomyopathy). Cardiac stem cell therapy poses three major challenges for researchers and clinicians: understand and improve immunosuppression and cell transplantation tolerance, understand the tissue-reproducing ability of specific cell types and their ability to effect autologous repair mechanisms, and evaluate new approaches to improve the efficacy of stem cell therapy particularly in the area of long-term cell survival.

Over 14,000 allogeneic UCB transplant procedures for hematologic malignancies have been performed internationally using UCB from related and unrelated donors in pediatric (3–6) and adult patients (7–10). The use of double cord transplants has shown considerable promise, possibly lowering the risk of graft rejection and the risk of relapse (10) where cell dose is directly related to successful outcomes (11). Even with a higher incidence of grade II acute graft versus host disease in recipients of two partially HLA-matched UCB units, there is no adverse effect on transplant-related mortality (12). The successful utilization of double umbilical cord blood grafts in thalassemia patients with early graft failure, with transplants separated by several months, was demonstrated (13), suggesting that a treatment regimen requiring several administrations of UCB retains effectiveness. Other strategies for successful clinical outcomes are based on expansion of the stem cells in CB grafts and the induction of more rapid engraftment with the modification of disease environments or of the stem cells themselves: these strategies will likely benefit approaches to regenerative medicine by enhancing the deposition of enhanced therapeutic concentrations of stem cells into the affected, diseased regions. Despite the reduced numbers of progenitor cells in UCB, collected data on unrelated cord blood transplants performed in European Blood and Marrow Transplant Group (EBMT) and non-EBMT centers show that the disease-free survival and overall survival of children with acute leukemia are similar to other hematopoietic stem cell sources (matched unrelated BM) (14).

Beyond curing malignant and nonmalignant diseases, cord blood has demonstrated preclinical efficacy. The ability of UCB to restore blood flow and mediate tissue repair mechanisms in preclinical models is well established and a number of clinical trials in process are exploring the ability of UCB to treat beyond hematological malignancies for type I diabetes, sickle cell anemia, and cerebral palsy. A number of preclinical studies have demonstrated the efficacy of human HSC transplantation in restoring blood flow and improving cardiac function in animal models of ischemia (15). Mechanisms by which these infused human cells mediate the improvement in blood and tissue regeneration remains unknown. Nonetheless, this preclinical laboratory work has initiated clinical trials utilizing individual patient-derived autologous bone marrow (BM) culture-derived cells or whole BM uncultured mononuclear cells (MNC) infused or injected locally in an attempt to augment angiogenesis in response to ischemia (16–20). Current sources of nonembryonic stem cells for these clinical trials are found in the bone marrow, mobilized peripheral blood (derived from the bone marrow), adipose tissue, and skeletal muscle but there is a need for randomized double-blind clinical trials using cord blood-derived stem cells to assess the overall efficacy of cell-based cord blood therapy. The idea that autologous bone-marrow derived stem cells (BMCs) can transdifferentiate into cardiomyocytes or vascular cells has been challenged in several scientific reports. The improvement in cardiac function after migration of autologous BMCs to the heart can be explained by their paracrine effects, inducing angiogenesis and preventing ischemic myocardium from apoptosis (21, 22). The heart has the ability to regenerate through activation of resident cardiac stem cells or through recruitment of a stem cell population from other tissues, such as bone marrow. Additional information has been obtained regarding new stem cell sources, cell-based gene therapy, cell-enhancement strategies, and tissue engineering, all of which should enhance the efficacy of human cardiac stem cell therapy.

Direct intramyocardial injection of stem cells into the myocardium is one route of delivery during surgical intervention.

This technique of local delivery of stem-progenitor cells to the myocardium has been shown to be feasible and safe in patients with heart disease. Other than open heart surgery, the intracoronary route appears to be the preferred approach in clinical studies because the stem cells are delivered directly to the affected area without further injury to the myocardium or risking possible negative systemic side effects of stem cell mobilization for the patient. One complementary approach to increase the efficiency of progenitor cell transplantation is to enhance cell recruitment and retention in the infarcted heart. For example, stromal cell-derived factor (SDF-1) has recently been shown to play a critical role in stem cell recruitment to bone marrow during transplantation for the treatment of leukemia and also to myocardial tissue after myocardial infarction. Although the underlying molecular mechanism is not clear, numerous studies in animals have shown that transplantation of bone marrow-derived stem cells or circulating endothelial progenitor cells following acute myocardial infarction and ischemic cardiomyopathy is associated with a reduction in infarction scar size and statistical improvements in left ventricular function and myocardial perfusion. The mechanism or mechanisms underlying marrow-derived HSC recruitment to ischemic regions are not well defined.

Although the current clinical approach of utilizing the patient-derived HSC has the advantage of avoiding potential adverse reactions to allogeneic cells by an immunocompetent host, there are distinctive disadvantages, including altered stem cell function and logistical issues related to cell collection from individual patients. A number of correlative studies to clinical trials have demonstrated significantly reduced efficacy of isolated stem cells to express critical chemokines receptors (e.g., CXCR4 and VEGF-R2), migrate to SDF-1, and augment blood flow in hindlimb and cardiovascular animal models. Therefore, an available ready source of allogeneic HSC such as umbilical cord blood (UCB) for cellular therapy may be optimal. One important caveat will be to address whether the intact immunogenicity of UCB-allogeneic HSC is advantageous to potentially augment angiogenesis by recipient cells in situ or potentially deleterious in dampening angiogenic mechanisms or responsible for worsening vascular ischemia by allogeneic inflammatory responses.

Sources of nonembryonic stem cells are located in the bone marrow, mobilized peripheral blood, adipose tissue, skeletal muscle, and umbilical cord blood. Cord blood contains multiple populations of stem cells identified by the expression of CD34+ and CD133+ surface antigens, a functional active marker of stem cells which contain aldehyde dehydrogenase and are classified as ALDH[bright] cells, the recent identification of very small embryonic stem cells (VSELs) (23), and mesenchymal stem cells (CD34 negative CD45 negative). These pluripotential stem cells are capable of giving rise to hematopoietic, epithelial, endothelial, and neural tissues both in vitro and in vivo. The identification and isolation of populations of pluripotent stem cells within cord blood represents a scientific breakthrough that could potentially impact every field of regenerative medicine. Thus, CB stem cells are readily utilized and amenable to treatment of hematological malignancies and also of a wide variety of nonmalignant diseases including cardiovascular, hepatic, ophthalmic, orthopedic, neurological, autoimmune, and endocrine diseases.

Adult HSC reside in the bone marrow but can be mobilized into the peripheral blood using exogenous G-CSF or GM-CSF. In addition, expression of CXCR-4, the receptor for stromal cell-derived growth factor (SDF-1), is expressed on the surface of CD34$^+$ cells (24). Following acute myocardial infarction (AMI), CD34$^+$ cells increase in the peripheral circulation (25, 26), occur rapidly following admission to the hospital, and then decline to control levels 60 days post-AMI. In heart failure patients, those with New York Heart Association (NYHA) Class I symptoms had increased numbers of circulating CD34$^+$ and CD34$^+$/CD133$^+$ cells: the number of circulating cells increase with NYHA class.

Expression of CXCR-4 on UCB CD34$^+$ cells is important in migration toward SDF-1-secreting stromal cells, in BM engraftment, and in cell engraftment in tissues secreting SDF-1 including myocardial tissue (27, 28). CB is an alternative source of stem cells with an inherent functional capability to augment angiogenesis which may provide therapeutic benefit beyond that of patient-derived autologous cells with demonstrated reduced functionality.

Other cell types, for example, human unrestricted somatic stem cells (USSC) have the potential to differentiate into myogenic cells and induce angiogenesis (29). The effect of USSC on myocardial regeneration and the improvement of heart function after acute myocardial infarction in a porcine model immunosuppressed by daily administration of cyclosporin A was demonstrated (30). Engrafted USSC were detected in the infarct region 4 weeks after cell transplantation and the implanted cells improved regional and global function of the porcine heart after a myocardial infarction. This study suggests that the USSC implantation is efficacious for cellular cardiomyoplasty but both the long-term follow-up and the requirement for sequential treatments with stem cells is unknown.

2. Materials

2.1. Protocols to Isolate and Characterize Umbilical Cord Stem Cells

1. Umbilical cord blood and bone marrow.
   (a) Ficoll (General Electric Healthcare).
   (b) AutoMACS and CliniMACS device devices (Miltenyi Biotec, Auburn, CA, USA).
2. 1% lidocaine.
   (a) EDTA (500 mM).
3. Antibodies: CD3-fluoroscein isothiocyanate (FITC), CD13-FITC, CD19-FITC (Invitrogen, Carlsbad, California), CD31-(phycoerythrin) PE, CD33-FITC, CD34-PE, CD45-FITC, CD56-FITC, CD61-FITC, CD105-FITC (Serotec, Raleigh, NC), CD133-PE (Miltenyi Biotec), CXCR4-APC (Allophycoerythrin), and KDR (VEGF-R2)-APC (R&D, Minneapolis, MN). All other antibodies were from Becton Dickinson, Franklin Lakes, NJ.
4. Colony-forming unit (CFU), granulocyte macrophage (GM), and burst-forming units-erthyroid (BFU-E) assay (StemCell Technologies, Vancouver, British Columbia).
5. LSR flow cytometer (Coulter, Miami, FL).
6. Compensation and data analysis were performed using WinList software (Verity Software House Inc., Topsham, ME).

3. Methods

3.1. UCB or BM CD133+/34+ Cell Isolation

Human umbilical cord blood and adult bone marrow from volunteer healthy donors were collected according to Institutional Review Board (IRB) protocols with informed consent. Mononuclear cells (MNC) were isolated by density gradient centrifugation on a Ficoll layer (15 ml) from freshly diluted UCB or BM (1:3 with phosphate buffered saline (PBS)), then CD133+ cells were isolated using magnetic separation as per the manufacturer's instructions for the research AutoMACS device. Alternatively for the clinical trial, collected bone marrow (175 ml) was diluted with PBS (1:5) and selected on a CliniMACS device.

3.2. Isolation of BM Cell Preparation and Delivery

For the clinical trial described in Adler et al., *Cell Transplantation* in press, autologous BM-derived CD133+ stem cell preparations were completed on the morning of the day of cardiac infusion.

1. BM aspiration was performed sterilely, the posterior iliac crest was prepped with betadine, and a 25-gauge needle introduced with 1% lidocaine to provide local anesthesia.

2. A 4-in., 11-gauge Jamsheedi biopsy needle was introduced into the marrow cavity and 150–200 ml removed. Following isolation of BM, mononuclear cells were isolated and selection of CD133⁺ cells was achieved by magnetic labeling of CD133⁺ cells using the CliniMACS device with CD133 microbeads.

3. Magnetically retained CD133⁺ cells were eluted as the positively selected fraction. The final cell harvest was centrifuged prior to final formulation and three washes in buffering solution were performed.

4. The final product was brought up to five ml in buffering solution at a concentration corresponding to the appropriate dose in PBS with 2 mM EDTA.

5. Sterility was assessed by gram stain and assays for mycoplasma and endotoxin as well as bacterial and fungal cultures. The minimum acceptable total number of infused CD133⁺ cells was 1×10^6 (range $0.5–1.5 \times 10^6$ cells), 2×10^6 (range $1.5–2.5 \times 10^6$ cells), and 3×10^6 (range $2.5–3.5 \times 10^6$ cells). A minimum purity and viability of 70% was deemed acceptable for infusion.

6. Mononuclear cells were assessed for colony-forming units per manufacturer protocol.

7. The cells were transported to the catheterization laboratory at ambient temperature for infusion within 8 h after final formulation.

3.3. Phenotypic Characterization and Assessment of CD133⁺ EPC Surface Phenotype and Purity by Flow Cytometry

Cellular aliquots were subjected to flow cytometry to evaluate the expression of cellular markers prior to and after cell selection. Several stem cell populations (e.g., CD34, CD133, mesenchymal stem cells, VSELs) have been isolated from cord blood. Previously, surface expressed receptors (e.g., CD133) in addition to the expression of CD34, have been identified on cells demonstrating hematopoietic reconstitution with an important role in the inter-relationship of stromal and hematopoiesis elements (31–33). Expression of CD34 is observed on umbilical vein endothelial cells but as these cells propagate through population doublings, CD34 surface expression is eventually lost (34). Cells isolated from UCB express CD34, VEGF-R2 (KDR), and CD133 and have the ability to migrate and differentiate into mature endothelial cells (35). Isolated cells cultured from peripheral blood from patients treated with granulocyte colony stimulating factor (G-CSF) in the presence of VEGF and stem cell growth factor (SCF) demonstrated expression of VEGF-R2, Tie-2, and von-Willebrand factor (vWF).

Cellular markers evaluated included PE-conjugated CD133 and FITC-conjugated CD45. Other lineage markers to determine other contaminating cells included CD3 (T-lymphocytes), CD13 (peripheral blood neutrophils, eosinophils, basophils, and monocytes), CD19 (B-lymphocyte), CD33 (myeloid marker), CD56 (NK cells), and CD61 (platelets). These surface phenotype analyses served to enumerate and characterize the purity of CD133+ cells and included identification and quantification of any contaminating cells.

1. After selection, cells were washed with PBS, counted, and viability was determined by trypan blue exclusion.

2. Surface phenotype was evaluated by incubation for 20 min at 4°C with fluorochrome-conjugated mAbs and appropriate isotype controls: CD31, CD34, CD105, CD133, CXCR4, and KDR (VEGF-R2).

3. Nonviable cells were excluded based on forward/side scatter and no additional gating was applied.

4. Positive staining quadrants were set based on nonstained samples and isotype controls. An LSR flow cytometer was used, acquiring >5,000 fluorescence events per sample.

5. Compensation and data analysis were performed using WinList software.

6. *Conclusions.* Ventricular remodeling determines the clinical progression of heart disease and is an important therapeutic target for the development of new medical and surgical treatment strategies. Over the last decade, cellular transplantation has emerged as an innovative biologic therapy that may restore myocardial structure and function. Utilizing current forms of cellular transplant therapies in clinical trials, confirmed myocardial regeneration remains limited and there is no therapy that has replaced the standard of care (36). Paracrine signals may provide beneficial effects by stimulating angiogenesis and vasculogenesis events, limiting extracellular matrix disruption, the onset and extension of fibrosis, and apoptosis. In addition, cell transplantation may directly induce endogenous cell mobilization and homing of endogenous repair mechanisms to injured myocardium through paracrine signals. Although complete myocardial regeneration remains the ultimate goal of cell therapy, the stimulatory abilities of cell transplantation can be exploited to complement current standard-of-care surgical approaches (e.g., percutaneous transluminal catheter angioplasty and stent placement) for patients with myocardial injury.

Previous clinical trials have demonstrated that injection or infusion into the myocardium or cardiovascular circulation, respectively, is safe and the use of autologous adult marrow or mobilized peripheral blood has advantages including the avoidance of potential adverse allogeneic immune reactivity. Nonetheless, there are disadvantages for routine clinical application of patient-derived stem cells, including the requirement for the mobilization of stem cell or bone marrow collection during active cardiovascular risk. In addition, the administration of cytokines to mobilize patient-derived peripheral blood stem cells may increase cardiovascular risk.

A number of angiogenesis techniques involving gene therapy and the use of growth factors have been explored in early clinical trials, but the resident, endogenous population of vascular endothelial cells in adults available to respond to angiogenic growth factors and chemokines limits the efficacy of these interventions due to age-related diminution of vascular endothelial cell number and function. In older patients, who are most likely to suffer from vascular problems, both centrally (i.e., coronaries) and peripherally, the number of hormone-responsive endothelial cells is reduced, and that of dysfunctional endothelial cells is increased (37, 38). This historical and more recent data support the concept that an exogenous, allogeneic source of HSC, rather than autologous patient-derived cells, may be optimal for cellular therapeutics intended to enhance angiogenesis, limit fibrosis and apoptosis, and facilitate cardiovascular function. UCB-derived HSC offer distinct advantages as a cell source, including greater potential lifespan and reparative proliferation relative to cells derived from peripheral blood or marrow of patients.

Early clinical trials using cell therapy attempting to augment the restorative process following acute or chronic myocardial ischemia have demonstrated some transient degree of efficacy. However, these clinical studies report variable efficacy and do not control for specified cell populations infused with therapeutic intent in a randomized, placebo-controlled manner. These clinical trial results indicate the importance of identifying the specific cell populations underlying the observed beneficial effects, and a need to conduct concomitant laboratory studies to gain an understanding of the mechanisms mediating augmentation of myocardial angiogenesis. There remains a need for further coordinated research with well-designed, hypothesis-driven clinical trials, in parallel with fundamental research aimed at understanding the mechanisms underlying the biological and functional effects of BM cell therapy for cardiac repair. In addition, key cell-specific features have not been addressed: patient-derived stem cells have reduced functional responsiveness and the comparison of the engraftment potential of young versus old mice has shown significant reductions in advanced aged mice indicating that mechanisms to home stem cells to bone marrow are defective. Recent studies have identified a second under-appreciated factor that the delivery of stem cells into the environment of advanced aged tissue shows reduced retention, with the local environment of aged tissue markedly defective in sustaining stem cells. This factor itself could potential limit efforts to facilitate tissue regeneration although a more robust stem cell source may counter this limitation.

The mechanism for these positive outcomes and the potential role of the immune system in mediating this process has yet to be elucidated. If the therapeutic benefit of infused HSC lies in augmentation of vasculogenesis via inflammatory signals elicited by stromal and hematopoietic cells present in situ, rather than direct anatomic localization of injected cells, this provides strong rationale for development of allogeneic HSC infusion with therapeutic vasculogenesis intent. Recently published trials using bone marrow-origin stem cells in cardiac repair reported a modest but significant benefit from this therapy but only after 4–6 months. Nonetheless, this initial benefit is not sustained with the longitudinal observation of patients. Further clinical research should aim to optimize the cell types utilized and their delivery mode and pinpoint optimal time of cell transplantation.

Effective therapies may capitalize on a treatment regimen using multiple stem cell populations delivered in concert with conventional surgical-based treatments (i.e., stent deployment or coronary artery bypass graft). Even in this case, patients may require multiple courses of stem cell treatment to maintain significant clinical efficacy and benefit.

4. Notes

1. Bone marrow aspirations were carried out without complication and aspirated volumes were adequate and similar among patients. Nonetheless, mononuclear cell preparations and subsequent selection of CD133+ cells were quite variable in number ranging from 8 to 48×10^6 (total CD133+ cells). These bone marrow aspirates yielded substantially more cells than required for infusion so future clinical trials using these infused cell concentrations could be decreased. Purity fell within the range of acceptable as defined in the protocol for all patients but was also highly variable (77–97% CD133+ CD45+). Similar highly variable results are observed with the research AutoMACS device and remain unexplained by the manufacturer.

2. In vitro analysis of the MNC preparations from all patients produced few and in some cases no colony-forming units. This observation may indicate that the suitability of a particular cell population considered for cellular therapy, especially when derived from a patient population, may require functional assessment in appropriate in vivo and/or in vivo assays prior to use.

3. All hematopoietic lineages were represented in the contaminating population, the largest percentage of which was myeloid precursors and megakaryocytes but no adverse reactions were noted during infusion of these autologous cells (Table 1). Therefore, there appears to be some latitude in obtaining an absolutely pure population of adult stem cells to use in these studies.

4.1. Angiogenesis

Defects in the regulation of new blood vessel growth (angiogenesis) or in vessel repair are major complications in cancer, diabetes, atherosclerosis, and myocardial infarction. In these diseases, the number of circulating endothelial progenitor cells (EPC) was altered and is associated with the angiogenic risk status and patient prognosis. The regulation of angiogenesis depends not only on the number of circulating EPC but also on their functional status. The initial paradigm underlying cellular therapy and how cells were hypothesized to exert reparative function to augment angiogenesis/regeneration or protection has involved the concept that transplanted cells would either differentiate into EC or myocardial cells in situ. However, this paradigm has been challenged by studies indicating that administered cells mediate angiogenesis via cell–cell interactions with injured vascular endothelium within the ischemic bed and by paracrine mechanisms. These cells home to injury sites, adhere to the activated and damaged endothelial cells or to the extracellular matrix and participate in the endothelial activation/repair process. Soluble factors VEGF and SDF-1 that potentiate angiogenesis and migration of stem cells are expressed in higher levels by EPC relative to mature EC (39). The concept of augmentation of angiogenesis by growth factor delivery to areas of ischemia is dependent on the transplanted populations of cells homing to injury sites (40, 41).

Chemokines regulate basal and inflammatory leukocyte trafficking and play a role in angiogenesis. Reperfused infarcted myocardium is associated with an inflammatory response resulting in leukocyte recruitment, healing, and fibrosis. Neutrophil chemoattractants (e.g., CXC chemokine CXCL8/interleukin (IL)-8) are up-regulated in the infarcted tissue inducing neutrophil infiltration. Mononuclear cell chemoattractants (e.g., CC chemokine CCL2/monocyte chemoattractant protein (MCP)-1) are expressed resulting in monocyte and lymphocyte recruitment into the ischemic area (42, 43). Another consideration is the activation state of transplanted cells which is critical for the vessel repair process. It may be possible to identify molecular targets crucial for EPC differentiation and function to test their involvement in EPC function during wound healing or tumor angiogenesis (44).

4.2. Cellular Differentiation and Trophic Mediation

Both circulating and transplanted stem cells home to ischemic tissue, but it is unclear to what extent these cells fuse and/or differentiate into other phenotypes. Preclinical studies with CB HSC

Table 1
CD133⁺ contamination characteristics

Patient	CD3	CD3⁻/CD56⁺	CD13⁺/SSC high	CD14	CD19 immature	CD19 mature	CD61	CD71
1	2.8	0.06	4.7	0.68	4.5	1.9	4.6	0.21
2	7.7	0.43	7.7	0.59	1.5	0.26	0.5	0.78
3	1.3	0.01	3.8	0.23	0.55	0.17	2.9	0.58
4	1.4	0.17	6.3	0.49	2.6	0.47	22.8	0.28
5	0.95	0.05	11.1	0.61	5.4	0.2	0.13	2.2
6	–	–	–	–	–	–	3.1	0.15
7	0.57	0.05	15.1	0.09	1.2	0.02	1.3	1.7
8	2.6	0.03	6.2	0.36	3.7	0.88	2.0	1.5
9	10.9	0.15	5.2	0.99	3.2	1.9	9.9	0.3
Summary	3.53 ± 3.50	0.12 ± 0.13	7.51 ± 3.55	0.51 ± 0.26	2.83 ± 1.58	0.73 ± 0.72	5.25 ± 6.79	0.86 ± 0.71

Absolute values are expressed as percent. Summary values are expressed as mean ± SD.

have evaluated whether direct incorporation of cells, differentiation of transplanted cells into either endothelial or cardiomyocytes, and/or paracrine are responsible for positive functional effects (45, 46). Studies differ in their conclusion regarding whether transplanted cells directly anatomically incorporate and differentiate into endothelial structures. Transplantation of CB HSC localize in the target tissue for a brief time period, although not quantitatively, and provide clear, functional, and structural improvement and a possible endogenous cell-survival benefit, that is, prevent apoptosis.

4.3. Involvement of the Innate Immunity

Several hypotheses have been postulated to explain the improved in vivo angiogenesis effects mediated by injected UCB HSC in murine models of vascular injury. Initially, the direct anatomic incorporation of the cells with or without the possibility of cellular differentiation has generated much debate. In contrast to this concept, many investigators have demonstrated that transplanted cells have the ability to indirectly promote recovery, preventing damage following reperfusion. This paracrine effect has the potential to be mediated in a number of different ways involving the immune system (47, 48). Innate immune responses to ischemia may be a compensatory mechanism that involves response to a pathological change in the environment (49). Mechanisms underlying inflammatory responses to acute vascular injury are not yet elucidated. Inhibition of complement in animal models can improve functional outcomes (50) and human clinical trials with anti-C5 complement antibody administration on patients with acute myocardium infarction has shown equivocal results (51). Migration of neutrophils and cellular infiltration into the myocardium is thought to promote and/or increase the degree of ischemia (52). Anti-inflammatory strategies to prevent infarction and promote recovery may not be effective and in fact may increase risk (53).

Inflammation plays an important role in healing via cellular infiltration and cytokine secretion. Monocytes have been shown to be instrumental regulators of the healing process in infarcted tissue (54). Our recent data demonstrate that patient-derived BM CD133+ stem cells used to treat chronic ischemic cardiac patients are defective in their constitutive expression of chemotactic cytokines IL8 and CCL5, or RANTES (Regulated upon Activation, Normal T-cell Expressed, and Secreted) is chemotactic for T lymphocytes, eosinophils, and basophils, and plays an active role in recruiting leukocytes into inflammatory sites. Thus immune mediation of ischemia-reperfusion is a balance of many factors. Modulating the balance toward a beneficial effect of the endogenous immune system may be one of the pathways with UCB HSC augment endothelial angiogenesis response to hypoxia.

4.4. Acute Versus Chronic (Ischemia) Myocardial Infarction

The therapeutic potential of human umbilical cord-derived stem cells in an acute rat myocardial infarction model demonstrates that transplanted cord blood cells provide benefit in cardiac function recovery (55). Histologic study and immunofluorescence were performed to investigate differentiation of transplanted cells, capillary and arteriole density, secretion of cytokines, and cardiomyocytes apoptosis. A statistically significant improvement of cardiac function was observed in the experimental group of rats compared with the control group. These results are supported by earlier observations that infarcted myocardium attracts cord blood mononuclear progenitor cells, that these cells substantially reduce myocardial infarction size and limit the myocardial concentration of tumor necrosis factor-alpha, monocyte/macrophage chemoattractant protein, monocyte inflammatory protein, and interferon-gamma in acutely infarcted myocardium (56). Reductions in these inflammatory proteins may indicate their participation in preventing the engagement of repair mechanisms.

4.5. Ex Vivo Delivered "Enhancing" Factors Influencing Stem Cell Homing, Retention

In human clinical trials, a modest but significant and sustained improvement in left ventricular function was observed in the *R*einfusion of *E*nriched *P*rogenitor Cells and *I*nfarct *R*emodeling in *A*cute *M*yocardial *I*nfarction (REPAIR-AMI) study contributing to the better clinical course. Results of other studies using bone marrow-derived total nucleated and stem cells were safe but only transiently effective at improving cardiovascular function. Differences in the study design, cell processing or timing of cell delivery might explain, in part, different outcomes among studies. Furthermore, studies in patients with chronic ischemic heart disease remain observational and conventional mechanisms to home stem cells (e.g., the SDF-1-CXCR4 interaction) does not appear to be operative.

A study has sought to optimize nonviral transfection of the human SDF-1 gene into skeletal myoblasts and transplant these cells to establish a transient SDF-1 gradient to favor extracardiac stem cell translocation into infarcted heart (57). These investigators concluded that ex vivo SDF-1 transgene delivery promotes stem and progenitor cell migration to the heart, activates cell survival signaling, and enhanced angiomyogenesis in the infarcted heart. However, recent studies have demonstrated arrhythmias associated with this transplantation strategy (58, 59).

Cord blood is a source of mesenchymal stem cells (60, 61) as well as hematopoietic stem cells and mesenchymal stem cells can differentiate into a number of cell types of mesenchymal lineage, such as cardiomyocytes, osteocytes, chondrocytes, and fat cells (62). Brown adipose tissue-derived cells differentiate into cardiomyocytes and these cardiomyocytes adapt functionally to repair regions of myocardial infarction. A novel approach to augment stem cell effectiveness is to coculture cord blood-derived mononuclear cells with brown adipose tissue-derived cells which leads to the most effective regeneration for impaired cardiomyocytes (63).

4.6. Early Clinical Trials

A number of early phase I/II cardiovascular cell-based therapy trials in humans demonstrated encouraging 4–6 month improvements in regional contractile function and ejection fraction comparing cell therapy-treated patients with standard therapy-treated patients (64, 65). However, further longitudinal follow-up studies (>1 year) demonstrated a loss of initially successful outcomes and limited improvements between groups (66) although analysis of pooled data indicates that BMC therapy in patients with acute and chronic cardiac issues results in modest improvements in left ventricular function and infarct scar size without negative effects (67). Several factors may have led to these results, including loss of incorporated treatment (i.e., stem cells) over time, choice of relatively healthy patient population with stable disease, and a placebo effect in nonrandomized and open-label studies. Nonetheless, any improvements may still be regarded as a successful outcome if further damage (increased apoptosis) and/or sustained damage (e.g., hypoxia) to myocardial tissue is averted by cell therapy.

4.7. Recent Clinical Trials (Stem Cells to Treat Damaged Myocardium)

Taken together, cell-based therapies provides an adjunct therapeutic approach to current revascularization surgical procedures and are safe, effective, and minimally invasive therapy to augment endogenous angiogenesis in both acute and chronic cardiovascular ischemia (68, 69). Although a number of preclinical studies have focused on using UCB stem cells to treat damaged myocardium (70, 71), there remains a pressing need for randomized double-blinded clinical trials using UCB-derived stem cells to assess the overall efficacy of cell-based cord blood therapy. In the use of cord blood to treat hematological malignancies, several enhancing mechanisms are being tested to increase the homing of stem cells to the bone marrow to hasten engraftment, a required initial step to repopulate the hematopoietic system. These mechanisms are targeted to increase the sensitivity of the CD34$^+$ CXCR4$^+$ expressing stem cells to SDF-1 from stromal cells: this identical mechanism appears to be operative in regenerative medicine applications involving the myocardium.

We drew a direct comparison between UCB and BM endothelial precursor cells (CD133$^+$) to determine their phenotypic differences and functional capacity to augment vasculogenesis (72). This study utilized an established NOD.SCID murine hindlimb injury model, which allows engraftment of human cells, where EPC derived from either BM or UCB were harvested and were injected into mice immediately after femoral artery ligation. Fourteen days following ligation, the ratios of ischemia to nonischemia blood flow were significantly higher in the injured leg in both study groups receiving stem cells but importantly there were no significant differences between the two sources of EPC indicating equivalent ability to augment angiogenesis. These preclinical results established the foundation to use BM-derived CD133⁺ in an attempt to augment cardiovascular circulation.

Therapies for coronary ischemia and subsequent myocardial necrosis include acute revascularization, coronary artery bypass grafting (CABG) and pharmacologic interventions targeted at reducing ischemia and ultimately preserving myocardium and prolonging life. Morbidity and mortality related to myocardial necrosis regardless of etiology have been reduced dramatically with the institution of these therapies. We continue to search for further ways in which to reduce the ischemic burden and ultimately facilitate improvement of symptoms. Specifically, patients with chronic total occlusion (CTO) provide a therapeutic challenge because they experience stable chronic angina which is difficult to control and or ameliorate (73). Percutaneous coronary intervention (PCI) to CTO account for approximately 12–16% of all interventions in the US (74). The Occluded Artery Trial recently revealed that PCI does not reduce MACE and may in fact provoke myocardial ischemia with increased reinfarction rates in clinically stable patients with persistent total occlusion (75). There are limited therapeutic options for patients with chronic ischemia secondary to CTO. Many of these patients have areas of viable but areas of myocardium at risk. Development of collateral growth is an important compensatory mechanism with respect to CTO. Collateralization may provide adequate augmentation of blood flow to ischemic tissue protecting patients with CTO from angina. However, collateralization may not provide this protection and further augmentation of collateral growth may be necessary to generate reduction or cessation of symptoms.

We postulated that a selected cell population may have a therapeutic advantage over whole cell preparations to provide a pure potent cell fraction for healing. Moreover, all current cellular studies have demonstrated safety with single dose applications without attempting to titrate the cell dose with respect to safety. Therefore, we aimed to determine if infusion with increasing cell dose of autologous bone marrow CD133⁺ selected stem cells was safe and feasible in patients with CTO.

A recent trial (Safety And Efficacy Of Autologous Intracoronary Stem Cell Injections In Total Coronary Artery Occlusions (SEACOAST trial)) demonstrated that intracoronary infusion of autologous CD133⁺ marrow-derived cells in a dose-escalating fashion is safe and feasible (76). This Phase I/II study was designed to assess the safety and feasibility of a dose escalating intracoronary infusion of autologous bone marrow (BM)-derived CD133⁺ stem cells in patients with CTO and ischemia and in long-term follow-up (to 1 year) to assess the efficacy of cell infusion. We observed that following stem cell infusion after 6 months, there were no major adverse cardiac events (cardiac and noncardiac) and no readmission to the hospital secondary to angina or acute myocardial infarction. There were no periprocedural infusion-related complications including but not limited to malignant arrhythmias, any angiographic evidence of loss of normal coronary blood flow or acute neurologic events. Cardiac enzymes to assess for myocardial damage were negative in all patients. There was an improvement in the degree of ischemic myocardium as assessed by nuclear stress imaging which was accompanied by a trend toward reduction in anginal symptoms at six month follow-up.

In this single center study of nine patients, the isolation and infusion of autologous CD133$^+$ cells was feasible. No in-hospital events were recorded including death, AMI, and ventricular arrhythmias. No patients were lost to follow-up and all clinical and ischemic functional assessments were completed. The primary outcome of safety over the course of the study was met. No MACE was incurred during the six-month follow-up period. There were no arrhythmias detected by 24-h Holter monitoring and no laboratory abnormalities attributable to infusion. With respect to secondary outcomes, seven of the nine patients had a reduction in the area of reference ischemia. There was suggestion of an increase in anginal stability and quality of life. However, there appeared to be no change in treatment satisfaction, anginal frequency, or improvement in physical limitation as assessed by the Seattle Angina Questionnaire.

In parallel to the clinical trial, we sought to determine whether significant differences in the biological potency of CD133$^+$ EPC exist between those cells derived from advanced aged patients with stable coronary ischemia as compared to those cells derived from both young, healthy controls and from CB. Both in vitro and in vivo studies comparing bone marrow-derived CD133$^+$ cells from patients with chronic coronary artery disease enrolled in the SEACOAST trial to those derived from both young, healthy controls and CB were performed (Section 4.1). We observed diminished biological potency of BM-derived CD133+ cells obtained from advanced aged patients with coronary artery disease with in vitro and in vivo studies (Greco et al., unpublished) (Section 4.2). Magnetically-selected CD133$^+$ cells from patient- and control-derived BM and from CB cells were evaluated for endothelial progenitor lineage by phenotyping and by endothelial cell colony formation (CFU-EC), tested functionally by in vitro transmigration, for their ability to augment blood flow in a murine hindlimb ischemia model, and for bone marrow homing and engraftment: all tested parameters were decreased with patient-derived cells relative to the control BM and CB groups. Transmigration of control- and patient-derived CD133$^+$ cells in response to 200 ng/ml stromal cell-derived factor-1 (SDF-1) was reduced 86% and transmigration to 50 ng/ml vascular endothelial growth factor (VEGF) was reduced 75%. To compare the ability of CD133$^+$ cells to home and engraft in the bone marrow, mice injected with 0.5×10^6 patient BM-derived CD133$^+$ cells versus those injected with control BM-derived CD133$^+$ cells showed <0.2% human CD45$^+$ cells versus $1.6 \pm 0.4\%$ human CD45$^+$ cells, respectively, in mice BM, whereas engraftment with CB-derived CD133$^+$ cell was 19%. Diminished biological potency of advanced aged patient-derived BM CD133$^+$ cells with advance age and persistent chronic myocardial ischemic was observed using both in vitro and in vivo studies when compared to umbilical cord blood and young (<40 years old) healthy donor-derived BM cells. Although we observed benefit of stem cell treatment for seven of nine patients, the long term benefit of utilizing patient-derived cells may be limited and suggest that an alternative adult stem cell such as cord blood stem cells may be a viable alternative. The limited positive clinical outcomes may be attributed to the local myocardial delivery and $1-3 \times 10^6$ CD133 stem cells. Previous trials have utilized a cell source which potentially may have reduced potential as a number of studies suggest that stem cell dysfunction may be inherent to patients with chronic disease (37, 38). Functional quantification was carried out in this study utilizing a CFU-EC with PB and BM MNC from each patient. None of the patient samples yielded significant numbers of colony-forming units compared to reduced aged controls suggesting a diminution of function.

4.8. Clinical Trials: Umbilical Cord Blood Hematopoietic Stem Cells

UCB contains pluripotent stem cells, are readily harvested after birth, and have clinical and logistical advantages over that of patient adult-derived stem cells in potential applications for therapeutic angiogenesis (77). Clinical advantages include UCB collection at no risk to the donor, greater accessibility following cryogenic storage, rapid availability in international accessible registries, availability of diverse HLA genotypes, lower T-lymphocyte immunoreactivity, and lower inherent pathogen transmission. The clinical use of UCB is not subject to the social and political controversy related to use of embryonic stem cells. UCB has been shown to be a source of HSC equivalent in cell density as bone marrow, recently embryonic-like stem cells have been identified in UCB (23) and UCB is the only tissue used for successful transplantation across HLA barriers (3). In transplantation, single UCB units can reconstitute the entire lympho-hematopoietic systems in both pediatric and adult patients (7) (8). UCB has higher proliferation capacity and longer telomeres (indicating more primitive cells) than equivalent aliquots of adult peripheral blood (PB) or BM (78, 79). Current clinical outcomes for leukemia transplantation indicates the requirement for UCB containing >1 billion cells to achieve rapid and durable engraftment to achieve a treatment dosing of $>2.5 \times 10^7$ TNC/kg of patient. Calculation of the dosing of CD34$^+$ cells indicates a need for $>10 \times 10^3$ CD34/kg body weight of patient. These clinical UCB are therefore sufficient to treat cardiovascular patients with $1–10 \times 10^6$ CD34/dose: all clinical trials to date utilize a single treatment with stem cells although multiple longitudinal treatments to sustain a biological effect may be required.

HLA-matched UCB HSC therefore may have distinct advantages as a cell source including greater potential cell lifespan and greater reparative proliferation, relative to existing models of therapeutic angiogenesis derived from patient PB or BM. Current in vivo animal results indicate that transplanted cells are not retained quantitatively in the injury site and the effects may be mostly paracrine.

Preclinical and clinical outcomes from the treatment of hematological malignancies have implications for regenerative medicine applications of cord blood-derived stem cells. Current attributes for the selection of UCB for malignant transplantation involves combining both cell dose and HLA matching as independent yet overlapping variables. For reliable engraftment, cell dose and cell yield are critical, given that the transplants are being performed with minimal cells. In transplants for malignant disorders, the greater allogenicity and lower relapse rate associated with the less well-matched units balance any benefit of better HLA matching on TRM (11). Although the overall conclusion is that UCB with greater cellular yields better outcomes, a better identification of contributing cells is still warranted. In the current use of cord blood in the treatment of hematological malignancies, substantial morbidity and mortality are associated with pretransplant ablation of the recipient hematopoietic system. Because of the unique immunological properties of cord blood, a possibility is to utilize allogeneic cells for regenerative applications without the need to fully compromise the recipient immune system by ablation (80). Besides the expansion of limited stem cells, changes in the expression of key homing receptors (e.g., CXCR4) may lead to increased deposition into myocardial tissue in a synonymous manner to increased engraftment in the bone marrow. Nevertheless, it remains unclear as to the efficacy dose of cord blood stem cells that may be required to treat ischemic myocardium.

The present consensus is that the beneficial effects of bone marrow-derived stem cells and other forms of cellular repair in post infarct patients occur independent of regeneration of cardiomyocyte formation. Insufficient endogenous homing mechanisms and survival of transplanted cells into the ischemic environment limit the full potential of cell-based cardiac repair. A complete understanding of the underlying molecular mechanisms of critical steps in injury and in cell-based repair will facilitate the development of improved clinical strategies to enhance functional recovery after myocardial infarction.

Additional stem cell populations in cord blood have been identified by functional responses (81). An on-going clinical trial (NCT00654433) will transplant eligible subjects with an unrelated umbilical cord blood transfusion as a possible cure for their inherited metabolic disease. ALDH^{bright} cells (ALD-101, Aldagen) will be separated from the cord blood unit and infused 4 h after standard cord blood transfusion. The primary outcome measure is to assess the efficacy of adjuvant therapy of ALD-101 in accelerating platelet engraftment in patients also receiving a standard unrelated UCBT for treatment of inherited metabolic diseases. These clinical trials will evaluate the safety of these populations and their ability to home to bone marrow. Their reparative activity in response to SDF-1 gradients may preclude their proposed use in treating cardiovascular diseases.

Despite the use of cord blood-derived cells in treating hematological malignancies, the use of this allogenic stem cell source in treating cardiovascular diseases has now been restricted to preclinical ischemic models. Transplantation with cord blood for malignancies requires pretransplant ablation of the recipient hematopoietic system. Because of the unique immunological properties of both the stem cell and nonstem cell components of cord blood, it may be possible to utilize allogeneic cells for regenerative applications without the need to compromise the recipient's immune system. The previous use of unmatched cord blood in the absence of any immune ablation, as well as potential steps for widespread clinical implementation of allogeneic cord blood grafts, may have direct implications for regenerative medicine. In this case, an HLA-*un*matched UCB HSC may have advantages in terms of the expected "rejection" by the patient but only after stimulation of endogenous repair mechanisms.

References

1. Watkins, L. O. (2004) Epidemiology and burden of cardiovascular disease. *Clin Cardiol* 27, III2–6.

2. Libby, P., Theroux, P. (2005) Pathophysiology of coronary artery disease. *Circulation* 111, 3481–8.

3. Rubinstein, P., Carrier, C., Scaradavou, A., et al. (1998) Outcomes among 562 recipients of placental-blood transplants from unrelated donors. *N Engl J Med* 339, 1565–77.

4. Kurtzberg, J. (1996) Umbilical cord blood: a novel alternative source of hematopoietic stem cells for bone marrow transplantation. *J Hematother* 5, 95–6.

5. Gluckman, E., Broxmeyer, H. A., Auerbach, A. D., et al. (1989) Hematopoietic reconstitution in a patient with Fanconi's anemia by means of umbilical-cord blood from an HLA-identical sibling. *N Engl J Med* 321, 1174–8.

6. Tse, W. W., Zang, S. L., Bunting, K. D., Laughlin, M. J. (2008) Umbilical cord blood transplantation in adult myeloid leukemia. *Bone Marrow Transplant* 41, 465–72.

7. Laughlin, M. J., Barker, J., Bambach, B., et al. (2001) Hematopoietic engraftment and survival in adult recipients of umbilical-cord blood from unrelated donors. *N Engl J Med* 344, 1815–22.

8. Laughlin, M. J., Eapen, M., Rubinstein, P., et al. (2004) Outcomes after transplantation of cord blood or bone marrow from unrelated donors in adults with leukemia. *N Engl J Med* **351**, 2265–75.

9. Atsuta, Y., Suzuki, R., Nagamura-Inoue, T., et al. (2009) Disease-specific analyses of unrelated cord blood transplantation compared with unrelated bone marrow transplantation in adult patients with acute leukemia. *Blood* **113**, 1631–8.

10. Appelbaum, F. R. (2008) Allogeneic hematopoietic cell transplantation for acute myeloid leukemia when a matched related donor is not available *Hematology Am Soc Hematol Educ Program* **2008**, 412–7.

11. Wall, D. A., Chan, K. W. (2008) Selection of cord blood unit(s) for transplantation. *Bone Marrow Transplant* **42**, 1–7.

12. MacMillan, M. L., Davies, S. M., Nelson, G. O., et al. (2008) Twenty years of unrelated donor bone marrow transplantation for pediatric acute leukemia facilitated by the National Marrow Donor Program. *Biol Blood Marrow Transplant* **14**, 16–22.

13. Jaing, T. H., Hung, I. J., Yang, C. P., Tsai, M. H., Lee, W. I., Sun, C. F. (2008) Second transplant with two unrelated cord blood units for early graft failure after cord blood transplantation for thalassemia. *Pediatr Transplant* 13(6):766–8.

14. Gluckman, E., Wagner, J. E. (2008) Hematopoietic stem cell transplantation in childhood inherited bone marrow failure syndrome. *Bone Marrow Transplant* **41**, 127–32.

15. Tepper, O. M., Capla, J. M., Galiano, R. D., et al. (2005) Adult vasculogenesis occurs through in situ recruitment, proliferation, and tubulization of circulating bone marrow-derived cells. *Blood* **105**, 1068–77.

16. Assmus, B., Honold, J., Schachinger, V., et al. (2006) Transcoronary transplantation of progenitor cells after myocardial infarction. *N Engl J Med* **355**, 1222–32.

17. Schachinger, V., Erbs, S., Elsasser, A., et al. (2006) Intracoronary bone marrow-derived progenitor cells in acute myocardial infarction. *N Engl J Med* **355**, 1210–21.

18. Schachinger, V., Erbs, S., Elsasser, A., et al. (2006) Improved clinical outcome after intracoronary administration of bone-marrow-derived progenitor cells in acute myocardial infarction: final 1-year results of the REPAIR-AMI trial. *Eur Heart J* **27**, 2775–83.

19. Schachinger, V., Tonn, T., Dimmeler, S., Zeiher, A. M. (2006) Bone-marrow-derived progenitor cell therapy in need of proof of concept: design of the REPAIR-AMI trial. *Nat Clin Pract Cardiovasc Med* **3** (Suppl 1), S23–8.

20. Lunde, K., Solheim, S., Aakhus, S., et al. (2006) Intracoronary injection of mononuclear bone marrow cells in acute myocardial infarction. *N Engl J Med* **355**, 1199–209.

21. Dzau, V. J., Gnecchi, M., Pachori, A. S., Morello, F., Melo, L. G. (2005) Therapeutic potential of endothelial progenitor cells in cardiovascular diseases. *Hypertension* **46**, 7–18.

22. Gnecchi, M., He, H., Noiseux, N., et al. (2006) Evidence supporting paracrine hypothesis for Akt-modified mesenchymal stem cell-mediated cardiac protection and functional improvement. *FASEB J* **20**, 661–9.

23. Kucia, M., Reca, R., Campbell, F. R., et al. (2006) A population of very small embryonic-like (VSEL) CXCR4(+)SSEA-1(+)Oct-4+ stem cells identified in adult bone marrow. *Leukemia* **20**, 857–69.

24. Aiuti, A., Turchetto, L., Cota, M., et al. (1999) Human CD34(+) cells express CXCR4 and its ligand stromal cell-derived factor-1. Implications for infection by T-cell tropic human immunodeficiency virus. *Blood* **94**, 62–73.

25. Kucia, M., Dawn, B., Hunt, G., et al. (2004) Cells expressing early cardiac markers reside in the bone marrow and are mobilized into the peripheral blood after myocardial infarction. *Circ Res* **95**, 1191–9.

26. Massa, M., Rosti, V., Ferrario, M., et al. (2005) Increased circulating hematopoietic and endothelial progenitor cells in the early phase of acute myocardial infarction. *Blood* **105**, 199–206.

27. Jo, D. Y., Rafii, S., Hamada, T., Moore, M. A. (2000) Chemotaxis of primitive hematopoietic cells in response to stromal cell-derived factor-1. *J Clin Invest* **105**, 101–11.

28. Askari, A. T., Unzek, S., Popovic, Z. B., et al. (2003) Effect of stromal-cell-derived factor 1 on stem-cell homing and tissue regeneration in ischaemic cardiomyopathy. *Lancet* **362**, 697–703.

29. Kogler, G., Sensken, S., Airey, J. A., et al. (2004) A new human somatic stem cell from placental cord blood with intrinsic pluripotent differentiation potential. *J Exp Med* **200**, 123–35.

30. Kim, B. O., Tian, H., Prasongsukarn, K., et al. (2005) Cell transplantation improves ventricular function after a myocardial infarction: a preclinical study of human unrestricted

somatic stem cells in a porcine model. *Circulation* **112**, I96–104.

31. Bonanno, G., Perillo, A., Rutella, S., et al. (2004) Clinical isolation and functional characterization of cord blood CD133+ hematopoietic progenitor cells. *Transfusion* **44**, 1087–97.

32. Civin, C. I., Almeida-Porada, G., Lee, M. J., Olweus, J., Terstappen, L. W., Zanjani, E. D. (1996) Sustained, retransplantable, multilineage engraftment of highly purified adult human bone marrow stem cells in vivo. *Blood* **88**, 4102–9.

33. Civin, C. I., Trischmann, T., Kadan, N. S., et al. (1996) Highly purified CD34-positive cells reconstitute hematopoiesis. *J Clin Oncol* **14**, 2224–33.

34. Fina, L., Molgaard, H. V., Robertson, D., et al. (1990) Expression of the CD34 gene in vascular endothelial cells. *Blood* **75**, 2417–26.

35. Peichev, M., Naiyer, A. J., Pereira, D., et al. (2000) Expression of VEGFR-2 and AC133 by circulating human CD34(+) cells identifies a population of functional endothelial precursors. *Blood* **95**, 952–8.

36. Fedak, P. W. (2008) Paracrine effects of cell transplantation: modifying ventricular remodeling in the failing heart. *Semin Thorac Cardiovasc Surg* **20**, 87–93.

37. Chauhan, A., More, R. S., Mullins, P. A., Taylor, G., Petch, C., Schofield, P. M. (1996) Aging-associated endothelial dysfunction in humans is reversed by l-arginine. *J Am Coll Cardiol* **28**, 1796–804.

38. Tschudi, M. R., Barton, M., Bersinger, N. A., et al. (1996) Effect of age on kinetics of nitric oxide release in rat aorta and pulmonary artery. *J Clin Invest* **98**, 899–905.

39. Urbich, C., Aicher, A., Heeschen, C., et al. (2005) Soluble factors released by endothelial progenitor cells promote migration of endothelial cells and cardiac resident progenitor cells. *J Mol Cell Cardiol* **39**, 733–42.

40. Tang, Y. L., Tang, Y., Zhang, Y. C., Qian, K., Shen, L., Phillips, M. I. (2005) Improved graft mesenchymal stem cell survival in ischemic heart with a hypoxia-regulated heme oxygenase-1 vector. *J Am Coll Cardiol* **46**, 1339–50.

41. Ikeda, Y., Fukuda, N., Wada, M., et al. (2004) Development of angiogenic cell and gene therapy by transplantation of umbilical cord blood with vascular endothelial growth factor gene. *Hypertens Res* **27**, 119–28.

42. Frangogiannis, N. G. (2004) Chemokines in the ischemic myocardium: from inflammation to fibrosis. *Inflamm Res* **53**, 585–95.

43. Frangogiannis, N. G. (2004) The role of the chemokines in myocardial ischemia and reperfusion. *Curr Vasc Pharmacol* **2**, 163–74.

44. Real, C., Caiado, F., Dias, S. (2008) Endothelial progenitors in vascular repair and angiogenesis: how many are needed and what to do? *Cardiovasc Hematol Disord Drug Targets* **8**, 185–93.

45. Kinnaird, T., Stabile, E., Epstein, S. E., Fuchs, S. (2003) Current perspectives in therapeutic myocardial angiogenesis. *J Interv Cardiol* **16**, 289–97.

46. Walter, D. H., Haendeler, J., Reinhold, J., et al. (2005) Impaired CXCR4 signaling contributes to the reduced neovascularization capacity of endothelial progenitor cells from patients with coronary artery disease. *Circ Res* **97**, 1142–51.

47. Pinhal-Enfield, G., Ramanathan, M., Hasko, G., et al. (2003) An angiogenic switch in macrophages involving synergy between Toll-like receptors 2, 4, 7, and 9 and adenosine A(2A) receptors. *Am J Pathol* **163**, 711–21.

48. Koczulla, R., Degenfeld, G., Kupatt, C., et al. (2003) An angiogenic role for the human peptide antibiotic LL-37/hCAP-18. *J Clin Invest* **111**, 1665–72.

49. Matzinger, P. (2002) The danger model: a renewed sense of self. *Science* **296**, 301–5.

50. Fu, J., Lin, G., Zeng, B., et al. (2006) Anti-ischemia/reperfusion of C1 inhibitor in myocardial cell injury via regulation of local myocardial C3 activity. *Biochem Biophys Res Commun* **350**, 162–8.

51. Mahaffey, K. W., Granger, C. B., Nicolau, J. C., et al. (2003) Effect of pexelizumab, an anti-C5 complement antibody, as adjunctive therapy to fibrinolysis in acute myocardial infarction: the COMPlement inhibition in myocardial infarction treated with thromboLYtics (COMPLY) trial. *Circulation* **108**, 1176–83.

52. Litt, M. R., Jeremy, R. W., Weisman, H. F., Winkelstein, J. A., Becker, L. C. (1989) Neutrophil depletion limited to reperfusion reduces myocardial infarct size after 90 minutes of ischemia. Evidence for neutrophil-mediated reperfusion injury. *Circulation* **80**, 1816–27.

53. Solomon, D. H., Schneeweiss, S., Glynn, R. J., et al. (2004) Relationship between selective cyclooxygenase-2 inhibitors and acute myocardial infarction in older adults. *Circulation* **109**, 2068–73.

54. Weihrauch, D., Arras, M., Zimmermann, R., Schaper, J. (1995) Importance of monocytes/macrophages and fibroblasts for healing of

micronecroses in porcine myocardium. *Mol Cell Biochem* **147**, 13–9.

55. Wu, K. H., Zhou, B., Mo, X. M., et al. (2007) Therapeutic potential of human umbilical cord-derived stem cells in ischemic diseases. *Transplant Proc* **39**, 1620–2.

56. Henning, R. J., Burgos, J. D., Ondrovic, L., Sanberg, P., Balis, J., Morgan, M. B. (2006) Human umbilical cord blood progenitor cells are attracted to infarcted myocardium and significantly reduce myocardial infarction size. *Cell Transplant* **15**, 647–58.

57. Elmadbouh, I., Haider, H., Jiang, S., Idris, N. M., Lu, G., Ashraf, M. (2007) Ex vivo delivered stromal cell-derived factor-1alpha promotes stem cell homing and induces angiomyogenesis in the infarcted myocardium. *J Mol Cell Cardiol* **42**, 792–803.

58. Smith, R. R., Barile, L., Messina, E., Marban, E. (2008) Stem cells in the heart: what's the buzz all about? Part 2: Arrhythmic risks and clinical studies. *Heart Rhythm* **5**, 880–7.

59. Gepstein, L. (2008) Electrophysiologic implications of myocardial stem cell therapies. *Heart Rhythm* **5**, S48–52.

60. Mihu, C. M., Mihu, D., Costin, N., Rus Ciuca, D., Susman, S., Ciortea, R. (2008) Isolation and characterization of stem cells from the placenta and the umbilical cord. *Rom J Morphol Embryol* **49**, 441–6.

61. Martins, A. A., Paiva, A., Morgado, J. M., Gomes, A., Pais, M. L. (2009) Quantification and immunophenotypic characterization of bone marrow and umbilical cord blood mesenchymal stem cells by multicolor flow cytometry. *Transplant Proc* **41**, 943–6.

62. Chamberlain, G., Fox, J., Ashton, B., Middleton, J. (2007) Concise review: mesenchymal stem cells: their phenotype, differentiation capacity, immunological features, and potential for homing. *Stem Cells* **25**, 2739–49.

63. Yamada, Y., Yokoyama, S., Wang, X. D., Fukuda, N., Takakura, N. (2007) Cardiac stem cells in brown adipose tissue express CD133 and induce bone marrow nonhematopoietic cells to differentiate into cardiomyocytes. *Stem Cells* **25**, 1326–33.

64. Stamm, C., Westphal, B., Kleine, H. D., et al. (2003) Autologous bone-marrow stem-cell transplantation for myocardial regeneration. *Lancet* **361**, 45–6.

65. Menasche, P., Hagege, A. A., Vilquin, J. T., et al. (2003) Autologous skeletal myoblast transplantation for severe postinfarction left ventricular dysfunction. *J Am Coll Cardiol* **41**, 1078–83.

66. Reffelmann, T., Konemann, S., Kloner, R. A. (2009) Promise of blood- and bone marrow-derived stem cell transplantation for functional cardiac repair: putting it in perspective with existing therapy. *J Am Coll Cardiol* **53**, 305–8.

67. Dawn, B., Abdel-Latif, A., Sanganalmath, S. K., Flaherty, M. P., Zuba-Surma, E. K. (2009) Cardiac repair with adult bone marrow-derived cells: the clinical evidence. *Antioxid Redox Signal* **11**, 1865–82.

68. Hill, J. M., Zalos, G., Halcox, J. P., et al. (2003) Circulating endothelial progenitor cells, vascular function, and cardiovascular risk. *N Engl J Med* **348**, 593–600.

69. Werner, N., Kosiol, S., Schiegl, T., et al. (2005) Circulating endothelial progenitor cells and cardiovascular outcomes. *N Engl J Med* **353**, 999–1007.

70. Furfaro, E. M., Gaballa, M. A. (2007) Do adult stem cells ameliorate the damaged myocardium? Human cord blood as a potential source of stem cells. *Curr Vasc Pharmacol* **5**, 27–44.

71. Jaing, T. H., Hsia, S. H., Chiu, C. H., Hou, J. W., Wang, C. J., Chow, R. (2008) Successful unrelated cord blood transplantation in a girl with malignant infantile osteopetrosis. *Chin Med J (Engl)* **121**, 1245–6.

72. Finney, M. R., Greco, N. J., Haynesworth, S. E., et al. (2006) Direct comparison of umbilical cord blood versus bone marrow-derived endothelial precursor cells in mediating neovascularization in response to vascular ischemia. *Biol Blood Marrow Transplant* **12,** 585–93.

73. Prasad, A., Rihal, C. S., Lennon, R. J., Wiste, H. J., Singh, M., Holmes, D. R., Jr. (2007) Trends in outcomes after percutaneous coronary intervention for chronic total occlusions: a 25-year experience from the Mayo Clinic. *J Am Coll Cardiol* **49**, 1611–8.

74. Stone, G. W., Gersh, B. J. (2006) Facilitated angioplasty: paradise lost. *Lancet* **367**, 543–6.

75. Hochman, J. S., Lamas, G. A., Buller, C. E., et al. (2006) Coronary intervention for persistent occlusion after myocardial infarction. *N Engl J Med* **355**, 2395–407.

76. Dale S. Adler, H. L., Ravi Nair, Jonathan L. Goldberg, Nicholas J. Greco, Tom Lassar, Daniel I. Simon, Mary J. Laughlin, and Vincent J. Pompili. (in press) Safety and efficacy of autologous intracoronary CD133+ stem cell injections in chronic total occlusions (SEACOAST trial). *Cell Transplant*.

77. Fredrickson, J. K. (1998) Umbilical cord blood stem cells: my body makes them, but do I get to keep them? Analysis of the FDA

proposed regulations and the impact on individual constitutional property rights. *J Contemp Health Law Policy* **14**, 477–502.

78. Nakahata, T., Ogawa, M. (1982) Hemopoietic colony-forming cells in umbilical cord blood with extensive capability to generate mono- and multipotential hemopoietic progenitors. *J Clin Invest* **70**, 1324–8.

79. Broxmeyer, H. E., Douglas, G. W., Hangoc, G., et al. (1989) Human umbilical cord blood as a potential source of transplantable hematopoietic stem/progenitor cells. *Proc Natl Acad Sci U S A* **86**, 3828–32.

80. Riordan, N. H., Chan, K., Marleau, A. M., Ichim, T. E. (2007) Cord blood in regenerative medicine: do we need immune suppression? *J Transl Med* **5**, 8.

81. Storms, R. W., Trujillo, A. P., Springer, J. B., et al. (1999) Isolation of primitive human hematopoietic progenitors on the basis of aldehyde dehydrogenase activity. *Proc Natl Acad Sci U S A* **96**, 9118–23.

Chapter 4

Isolation of Resident Cardiac Progenitor Cells by Hoechst 33342 Staining

Otmar Pfister, Angelos Oikonomopoulos, Konstantina-Ioanna Sereti, and Ronglih Liao

Abstract

Cardiac resident stem/progenitor cells are critical to the cellular and functional integrity of the heart by maintaining myocardial cell homeostasis. Given their central role in myocardial biology, resident cardiac progenitor cells have become a major focus in cardiovascular research. Identification of putative cardiac progenitor cells within the myocardium is largely based on the presence or absence of specific cell surface markers. Additional purification strategies take advantage of the ability of stem cells to efficiently efflux vital dyes such as Hoechst 33342. During fluoresence activated cell sorting (FACS) such Hoechst-extruding cells appear to the side of Hoechst-dye retaining cells and have thus been termed side population (SP) cells. We have shown that cardiac SP cells that express stem cell antigen 1 (Sca-1) but not CD31 are cardiomyogenic, and thus represent a putative cardiac progenitor cell population. This chapter describes the methodology for the isolation of resident cardiac progenitor cells utilizing the SP phenotype combined with stem cell surface markers.

Key words: Hoechst 33342, FACS, Side population (SP), Cardiac progenitor cells

1. Introduction

The identification of resident cardiac stem cells with the potential to differentiate into multiple cardiac cell lineages has added an important new dimension to myocardial biology. In order to properly study cardiac stem cells, reliable methods for the identification and isolation of putative cardiac stem cells are crucial. Over the last 5 years several isolation protocols have been published (1–5). Most of these protocols rely on the presence of stem-cell associated cell surface markers including stem cell antigen 1 (Sca-1), the receptor for cytokine stem cell factor, steel factor (c-kit), as well as the ATP-binding cassette (ABC) transporters

Randall J. Lee (ed.), *Stem Cells for Myocardial Regeneration: Methods and Protocols*, Methods in Molecular Biology, vol. 660, DOI 10.1007/978-1-60761-705-1_4, © Springer Science+Business Media, LLC 2010

Mdr1 and Abcg2, and the absence of cell-lineage markers, including lineage, CD45, and CD31, among others (2, 4–7). While enzymatic tissue digestion potentially alters the expression of cell surface markers via cleavage (e.g., c-kit), ABC-transporter activity is not affected by the digestion process (5).

Functionally, ABC-transporter activity can be conveniently and reliably assessed by challenging cells with the DNA-binding dye Hoechst 33342 (8, 9). Upon Hoechst 33342 exposure, ABC-transporter-competent cells efficiently clear the fluorescent dye Hoechst 33342, thereby becoming "Hoechst-low." Putative stem cells are thus identified as the cellular fraction with the lowest Hoechst fluorescence intensity during fluorescence-activated cell sorting (FACS) (8). As Hoechst-low cells characteristically appear to the side of Hoechst dye retaining cells in the FACS profile, Hoechst-low cells have traditionally been termed side population (SP) cells. ABC-transporter activity is ATP and calcium dependent, thus, blocking of the ATP-binding site by the calcium channel blocker verapamil abolishes the Hoechst-efflux phenomenon. Hence, verapamil is commonly used to document the specificity of the SP pattern during FACS analysis (10).

Utilizing Hoechst 33342 dye staining to identify mononuclear cells containing stem cell activity was first introduced by Goodell et al. (8). In their seminal work, the authors demonstrated that Hoechst-extruding SP cells isolated from bone marrow are enriched in hematopoietic stem cell properties. Subsequently, the SP phenotype has been used to isolate stem/progenitor cells successfully from various solid tissues (10). We have modified the protocol by Goodell et al. and established a protocol exclusively for the isolation of SP cells from adult myocardium (5). The myocardium contains a subset of cardiac progenitor cells that express the ABC transporters Abcg2 and Mdr1 necessary for Hoechst-extrusion (Fig. 1). In fact, the Hoechst efflux ability, i.e., SP phenotype, of cardiac SP cells is regulated by both Abcg2 and Mdr1 in an age-dependent fashion (11). Thus, developmental status and age might significantly impact the yield and phenotype of cardiac SP cells. To further isolate the stem/progenitors with higher cardiomyogenic potential, we have combined SP isolation techniques with immunostaining for Sca-1 and CD31 to select for Sca-1+/CD31– cardiac SP, a sub population of cardiac SP cells with enriched cardiomyogenic potential (5).

2. Materials

2.1. Enzymatic Digestion of Murine Hearts

1. Collagenase B, lyophilizate (Roche Applied Science). The final concentration for the digestion is 0.1%. Store dry at 4°C and protected from light.

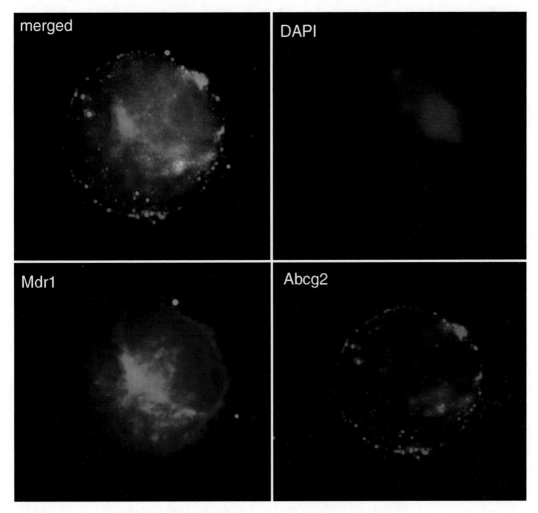

Fig. 1. Cardiac SP cells expressing Abcg2 and Mdr1. Freshly isolated cardiac SP cells stained for Mdr1 (*green*) and Abcg2 (*red*). The nuclei were stained with DAPI (*blue*).

2. Dispase II, lyophilizate (Roche Applied Science). The final concentration for the digestion is 2.4 U/ml in 5.56 mM glucose, 138 mM NaCl, 5.4 mM KCl, and 4.16 mM NaHCO$_3$.

3. CaCl$_2$. The final concentration for the digestion is 2.5 mM.

4. Hanks Balanced Salt Solution (HBSS) buffer: HBSS (Invitrogen) supplemented with 2% fetal bovine serum (GIBCO) and 10 mM HEPES.

2.2. Hoechst 33342 Incubation

1. Hoechst 33342 (Sigma Aldrich) dissolved in water at a final concentration of 1 mg/ml, filter sterilized.

2. High glucose Dulbecco's Modified Eagle's Medium (DMEM) buffer: DMEM (Cellgro) supplemented with 2% fetal bovine serum (GIBCO) and 10 mM HEPES.

3. Verapamil (Sigma) 5 mM stock solution in 95% ethanol.

2.3. Cell Surface Antigen Staining

1. HBSS buffer: Hanks Balanced Salt Solution (HBSS) (Invitrogen) supplemented with 2% fetal bovine serum (GIBCO) and 10 mM HEPES.

2. Sca-1 antibody: Phycoerythrin (PE) conjugated rat antimouse (BD Pharmingen). Isotype control PE Rat IgG2a, κ (BD Pharmingen).

3. CD31 antibody: Allophycocyanin (APC) conjugated rat antimouse (BD Pharmingen). Isotype control APC Rat IgG2a, κ (BD Pharmingen).

4. Flow cytometers: MoFlo (Cytomation, Inc). FACSAria (BD Bioscience).

2.4. FACS Analysis and Sorting

1. FACS tubes (BD Pharmingen).

2. Filters (40 μm and 70 μm, BD Pharmingen).

3. Propidium iodide (PI) (Sigma) is dissolved in water at 200 μg/ml and stored in small aliquots at –20°C. PI is used at a final concentration of 2 μg/ml.

3. Methods

This Hoechst SP cell purification protocol was established for the isolation of murine cardiac SP progenitor cells from C57Bl/6 and FVB mice. It is important to note that total cardiac SP cell yield, as well as the proportion of Sca-1+/CD31– cardiac SP cells, is age dependent with higher SP cell numbers in neonatal hearts compared to adult hearts (5). According to our experience with the aforementioned mouse strains, an average of 0.8 ± 0.2% cardiac SP cells can be expected in mice 10–12 weeks of age.

As the ability to discriminate Hoechst low SP cells is based on an active biological process, namely the energy dependent efflux of Hoechst 33342 by the ABC-transporters Abcg2 and Mdr1 (11), careful attention must be applied to the staining conditions in order to obtain an optimal resolution for the FACS profile. Correct Hoechst concentration, accurate cell counting, exact staining time, and staining temperature are critical. In order to prohibit further dye efflux after the staining process, it is crucial to keep the cells at 4°C until the FACS analysis is performed.

3.1. Enzymatic Digestion of Murine Hearts

1. Make fresh digestion buffer containing 0.1% Collagenase B, 2.4 U/ml Dispase II, $CaCl_2$ 2.5 mM. Prewarm this digestion buffer in a 37°C water bath prior to use. Utilize 5 ml digestion buffer per mouse-heart. For pooled hearts digestion, calculate the total amount of digestion buffer accordingly (Note that for optimal digestion, it is best to pool no more than four hearts).

2. Anesthetize the mouse with pentobarbital (65 mg/kg body weight i.p.). Note that alternative methods of anesthesia can be used in accordance with the individual Institutional Animal Care and Use Committee's guidelines and approval. Once the mouse is fully anesthetized, wipe the chest region gently with 70% ethanol. Cut the rib cage to expose the thoracic cavity. Open the thorax to expose the heart and flush the heart with 10 ml PBS. Use forceps to lift and cut the heart out and then place the heart immediately on ice in a culture dish containing cold PBS to wash away any residual blood.

3. After washing, remove all PBS and place the hearts in a 60×15 mm culture dish. Trim away the great vessels and atria. Cut the hearts with sterile scissors into small pieces and further mince with a sterile razor blade. Add 5 ml digestion buffer per heart for single heart digestion or 20 ml per four hearts for pooled hearts digestion. Thoroughly homogenate the minced cardiac tissue in digestion buffer by passing through a 2 ml pasteur pipette multiple times. Incubate at 37°C for 30 min and ensure that the tissue is entirely submerged in the digestion buffer. During the 30 min incubation time, pass the tissue homogenate through the pasteur pipette one more time about 15 min into the incubation period in order to disrupt connective tissue.

4. Following 30 min of incubation, stop the enzymatic reaction by diluting the digestion buffer with equal volumes of cold HBSS buffer, and filter through a 70 μm filter. Centrifuge at 530 g for 5 min at 4°C and carefully remove the supernatant. Resuspend the cells in the same HBSS buffer as described previously followed by filtering cells again through a 40 μm filter. Take 10–20 μl of this cell suspension and count the mononucleated cells with a haemocytometer. Note that cell counting is the most crucial step of the entire isolation procedure and one must be sure to not include erythrocytes in the total cell count. Any counting errors that might occur in the cell counting will lead to the incorrect Hoechst dye/total cell number ratio in the subsequent Hoechst staining. We find an average number of $3–5×10^6$ cardiomyocyte-depleted mononucleated cells per heart from a ~10-week-old C57Bl/6 mouse.

3.2. Hoechst Incubation

As the Hoechst staining procedure is temperature sensitive, it is of utmost importance to ensure that the water bath is set precisely at 37°C. Temperature fluctuations during incubation should be avoided if possible.

1. Centrifuge the cell suspension at 530 g for 5 min at 4°C and carefully remove the supernatant. Resuspend the cells at 10^6 cardiomyocyte- and erythrocyte-depleted (Notes 1 and 2)

mononucleated cells per ml in prewarmed (37°C) DMEM buffer (DMEM buffer refers to the DMEM media containing FBS and HEPES as described in Subheading 2.2 above) and add Hoechst 33342 to a final concentration of 1 μg/ml (Note 3). Mix the cells and Hoechst dye well and place the tubes in a 37°C water bath protected from light for exactly 90 min. It is important to ensure that the temperature in each tube is maintained precisely at 37°C. During the 90-min incubation time period, gently shake the tubes in order to remix the cell suspension every 15 min.

2. Always prepare a negative control sample by including verapamil (50 μM) during the entire Hoechst staining procedure. Verapamil blocks the ABC-transporter dependent Hoechst dye efflux and thus will serve to distinguish Hoechst retaining cells from SP cells upon FACS analysis. Because this negative control sample determines the threshold of the respective SP gate, it is an essential element of the SP analysis and should be included in each staining protocol.

3. After the 90 min incubation time, stop the reaction by placing the tubes on ice and spin the cells down at 530 g for 5 min at 4°C. Remove the supernatant and resuspend cells in cold HBSS buffer. Wash cells twice with cold HBSS buffer. At this point, samples are ready for FACS analysis or for surface marker staining (for details see the section below) for additional subpopulation selection or analysis. It is important to note that from this point onward all further manipulations should be performed at 4°C (on ice).

3.3. Cell Surface Antigen Staining

After the last washing step (step 3 in the section described above), resuspend cells at 10^6 cells per 100 μl cold HBSS buffer and stain with PE-conjugated rat antimouse antibody reactive to Sca-1 and APC-conjugated rat antimouse antibody reactive to CD31 at a concentration of 1/100. Incubate samples in the dark for 30 min at 4°C. After incubation, wash cells twice in cold HBSS buffer and proceed with FACS analysis. Propidium iodide (PI) (2 μg/ml) or 7-AAD (7-Amino-actinomycin D) (0.25 μg/10^6 cells) is added to each sample prior to FACS analysis to exclude dead cells.

3.4. FACS Analysis

FACS can be performed in any commercially available flow cytometer equipped with ultraviolet (UV) laser with the capacity to excite at 350 nm and to collect emissions at 450 and 650 nm. We have satisfactory experience with several different flow cytometers including MoFlo and FACSAria equipped with triple lasers. The availability of an UV laser to excite the Hoechst dye is a prerequisite for performing Hoechst dye efflux analysis. Hoechst dye is excitable at the wavelength of 350 nm (UV range) and its dual emission fluorescence is measured at 450–480 nm (Hoechst blue) and 650–680 nm (Hoechst red). A long pass dichroic mirror is used to separate the emission wavelengths. Note that the precise excitation and emission wavelengths may differ slightly depending on the setup of the given instrumentation, though it should fall within the excitation and emission spectrum of Hoechst 33342 dye. It is recommended to consult with an experienced FACS operator and the user manual of each flow cytometer (Note 4).

For those with limited experience in FACS and/or SP cell isolation, it is recommended to perform dose-dependent studies with various concentrations of Hoechst 33342 dye to cell number ratio. This dose-dependent experiment in combination with verapamil will help to determine the optimal concentration of Hoechst dye per given cell number. During the analysis and/or sorting, it is recommended that samples be kept on ice to maintain the quality of the specimen. Moreover, PI or 7-AAD positive cells represent nonviable cells, and should be excluded from the subsequent analysis by proper gating. As showed in Fig. 2, the percentage of Hoechst low or negative cells increases as the Hoechst concentration is lowered (Fig. 2a–c). This phenomenon is due to intrinsic Hoechst efflux abilities of ABC-transporters. When the Hoechst concentration is much lower than the capacity of ABC-transporters, Hoechst low cells or SP profile is artificially increased, whereas, when the Hoechst concentration exceeds the efflux capacity of ABC-transporters, the cells appearing as Hoechst low decrease drastically, leading to an incorrect SP profile. As such, Hoechst dose-dependent studies in combination with verapamil will aid in determining an optimal Hoechst/cell number ratio. Using this method, we find that 1 µg/ml Hoechst is the best concentration (Fig. 2b) to obtain a SP profile that can completely be inhibited by verapamil (Fig. 2e), with a lower or higher Hoechst concentration leading to understaining and overstaining, respectively.

Once the optimal Hoechst concentration has been selected, SP cells may be obtained as described below.

1. Nonviable or dead cells are excluded by displaying PI or 7-AAD (vertical axis) versus forward scatter parameters. Then, the Hoechst Blue versus Red profile is displayed, with Blue on the vertical axis (405 BP filter) and Red (660 LP filter) on the horizontal axis on linear scale. For voltage adjustments, it is important to put detectors in linear mode. Voltages are adjusted in a way that contaminating erythrocytes are located in the lower left corner and the bulk of nonerythrocytes are centered (Fig. 3). Once such a profile is displayed, draw a gate excluding erythrocytes and display the gate in a new window. For cardiac SP analysis, we recommend to collect

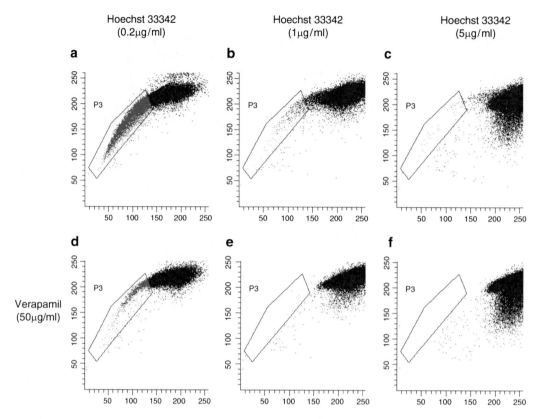

Fig. 2. (**a–e**) Titration of the optimal Hoechst/cell number ratio. Characteristic FACS dot-plot of Hoechst-stained cardiac cells with Hoechst red fluorescence on the horizontal and Hoechst blue fluorescence on the vertical axis. The boxed area depicts the region of Hoechst-low side population (SP) cells. The number of Hoechst-extruding SP cells within the boxed area is directly dependent on the Hoechst-concentration (Hoechst/cell number ratio). An inappropriately low Hoechst/cell number ratio (understaining) leads to an excess in SP cells (**a**) because of nonspecificity as demonstrated by the incomplete inhibition of the SP phenotype after coincubation with the ABC-transporters inhibitor verapamil (**d**). An inappropriately high Hoechst/cell number ratio (overstaining) overwhelms the ability of ABC-transporters and thus, dramatically reduces the percentage of SP cells (**c** + **f**). The optimal Hoechst/cell number ratio provides a robust SP population (**b**) that is entirely inhibited after coincubation with verapamil (**e**).

a minimum of 100,000 events within this gate. With optimal resolution, the cardiac SP profile should appear as shown in Fig. 2b, with SP cells located in the boxed area.

2. Confirm the specificity of the SP profile using the negative control (sample coincubated with verapamil and Hoechst). In this sample, no cells should appear in the boxed area (Fig. 2e).

3. After confirming the right position of the SP gate, sort SP cells in desired media. The SP population can be further purified or analyzed by staining with various surface markers. PE-conjugated rat antimouse antibody reactive to Sca-1 and APC-conjugated rat antimouse antibody reactive to CD31 at a concentration of 1/100 have previously been used by our laboratory.

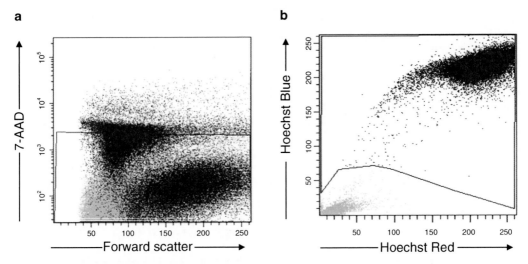

Fig. 3. Separation of Hoechst-stained cardiac cells from nonstained erythrocytes. For separation of Hoechst-stained cardiac cells from Hoechst-negative erythrocytes voltage of FACS detectors should be adjusted in linear mode in a way that contaminating erythrocytes are located in the lower left corner and the bulk of nonerythrocytes (Hoechst-stained cardiac cells) are centered.

Incubate samples in the dark for 30 min at 4°C. FACS analysis is preformed following the incubation and washout. Within the SP gate (boxed area), one can further display the expression of Sca-1 and CD31, alone or in combination, in a new dot plot with APC on the vertical axis and PE on the horizontal axis as shown in Fig. 4.

4. Notes

1. In our hands, the combination of collagenase B and dispase II reliably digests all mature cardiomyocytes, thus the resulting cardiac cell suspension is free from contaminating cardiomyocytes and no additional purification is needed. Other groups have used different digestion protocols, using pronase (2) for tissue digestion followed by a 30–70% Percoll gradient to remove residual cardiomyocytes to obtain cardiomyocyte-free cardiac cell suspensions for cardiac SP cell analysis.

2. Red blood cell lysis buffer or Ficoll gradient can be used to remove excess red blood cells.

3. As cell counting may vary quite a bit among users, 1 µg/ml Hoechst should not be the generalized concentration for everyone. Therefore, it is necessary to perform Hoechst dose-dependent curves in combination with verapamil to

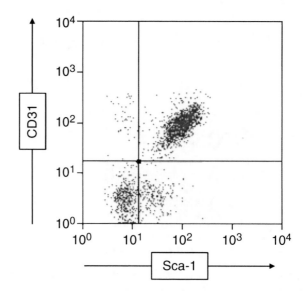

Fig. 4. Further purification by cell surface antigen staining. Subpopulations of cardiac SP cells according to Sca-1/CD31 staining. FACS analysis of cardiac SP cells following Sca-1 and CD31 labeling demonstrating that Sca-1+/CD31+ represent the majority, and Sca-1+/CD31− the minority of total cardiac SP cells.

determine the optimal Hoechst/cell number ratio per ml solution.

4. For the best results, it is recommended to work with low differential pressures and low velocities during FACS analysis.

Acknowledgments

This work was supported by NIH grants HL71775, HL86967, HL73756, HL 88533, and HL93148. The authors would like to thank Mr. Darragh Cullen for technical assistance. Mr. Grigoriy Losyev at BWH Cardiovascular FACS Core is acknowledged for assistance with cell sorting.

References

1. Laugwitz, K. L., Moretti, A., Lam, J., Gruber, P., Chen, Y., Woodard, S., Lin, L. Z., Cai, C. L., Lu, M. M., Reth, M., Platoshyn, O., Yuan, J. X., Evans, S., and Chien, K. R. (2005) Postnatal isl1+ cardioblasts enter fully differentiated cardiomyocyte lineages, *Nature 433*, 647–653.

2. Martin, C. M., Meeson, A. P., Robertson, S. M., Hawke, T. J., Richardson, J. A., Bates, S., Goetsch, S. C., Gallardo, T. D., and Garry, D. J. (2004) Persistent expression of the ATP-binding cassette transporter, Abcg2, identifies cardiac SP cells in the developing and adult heart, *Dev Biol 265*, 262–275.

3. Messina, E., De Angelis, L., Frati, G., Morrone, S., Chimenti, S., Fiordaliso, F., Salio, M., Battaglia, M., Latronico, M. V., Coletta, M., Vivarelli, E., Frati, L., Cossu, G., and Giacomello, A. (2004) Isolation and expansion of adult cardiac stem cells from human and murine heart, *Circ Res 95*, 911–921.

4. Oh, H., Bradfute, S. B., Gallardo, T. D., Nakamura, T., Gaussin, V., Mishina, Y.,

Pocius, J., Michael, L. H., Behringer, R. R., Garry, D. J., Entman, M. L., and Schneider, M. D. (2003) Cardiac progenitor cells from adult myocardium: homing, differentiation, and fusion after infarction, *Proc Natl Acad Sci U S A 100*, 12313–12318.

5. Pfister, O., Mouquet, F., Jain, M., Summer, R., Helmes, M., Fine, A., Colucci, W. S., and Liao, R. (2005) CD31⁻ but Not CD31⁺ cardiac side population cells exhibit functional cardiomyogenic differentiation, *Circ Res 97*, 52–61.

6. Beltrami, A. P., Barlucchi, L., Torella, D., Baker, M., Limana, F., Chimenti, S., Kasahara, H., Rota, M., Musso, E., Urbanek, K., Leri, A., Kajstura, J., Nadal-Ginard, B., and Anversa, P. (2003) Adult cardiac stem cells are multipotent and support myocardial regeneration, *Cell 114*, 763–776.

7. Hierlihy, A. M., Seale, P., Lobe, C. G., Rudnicki, M. A., and Megeney, L. A. (2002) The postnatal heart contains a myocardial stem cell population, *FEBS Lett 530*, 239–243.

8. Goodell, M. A., Brose, K., Paradis, G., Conner, A. S., and Mulligan, R. C. (1996) Isolation and functional properties of murine hematopoietic stem cells that are replicating in vivo, *J Exp Med 183*, 1797–1806.

9. Zhou, S., Schuetz, J. D., Bunting, K. D., Colapietro, A. M., Sampath, J., Morris, J. J., Lagutina, I., Grosveld, G. C., Osawa, M., Nakauchi, H., and Sorrentino, B. P. (2001) The ABC transporter Bcrp1/ABCG2 is expressed in a wide variety of stem cells and is a molecular determinant of the side-population phenotype, *Nat Med 7*, 1028–1034.

10. Challen, G. A., and Little, M. H. (2006) A side order of stem cells: the SP phenotype, *Stem Cells 24*, 3–12.

11. Pfister, O., Oikonomopoulos, A., Sereti, K. I., Sohn, R. L., Cullen, D., Fine, G. C., Mouquet, F., Westerman, K., and Liao, R. (2008) Role of the ATP-binding cassette transporter Abcg2 in the phenotype and function of cardiac side population cells, *Circ Res 103*, 825–835.

Chapter 5

Mesenchymal Stem Cell Therapy for Cardiac Repair

Andrew J. Boyle, Ian K. McNiece, and Joshua M. Hare

Abstract

Stem cell therapy for repair of damaged cardiac tissue is an attractive option to improve the health of the growing number of heart failure patients. Mesenchymal stem cells (MSCs) possess unique properties that may make them a better option for cardiac repair than other cell types. Unlike other adult stem cells, they appear to escape allorecognition by the immune system and they have immune-modulating properties, thus making it possible to consider them for use as an allogeneic cell therapy product. There is a large and growing body of preclinical and early clinical experience with MSC therapy that shows great promise in realizing the potential of stem cell therapy to effect repair of damaged cardiac tissue. This review discusses the mechanism of action of MSC therapy and summarizes the current literature in the field.

Key words: Mesenchymal stem cell, Heart failure, Myocardial infarction, Cell therapy

1. Introduction

Stem cell therapy holds immense promise for repairing damaged cardiac tissue and treating heart failure. However, the optimal type of stem cell for this purpose remains unknown. Mesenchymal stem cells (MSCs) have several important properties that make them attractive as a possible cell source for cardiac repair. They can be easily accessed, they have a proven safety track record, they can be expanded in culture, and they are immune privileged, allowing them to be used as an allogeneic product. A vast amount of laboratory and preclinical data has demonstrated the safety and efficacy of MSC therapy for the repair of damaged myocardium, and we are appropriately poised to enter large-scale clinical trials of this therapy. In this chapter, we review the evidence for MSC therapy in cardiac repair.

Randall J. Lee (ed.), *Stem Cells for Myocardial Regeneration: Methods and Protocols*, Methods in Molecular Biology, vol. 660, DOI 10.1007/978-1-60761-705-1_5, © Springer Science+Business Media, LLC 2010

1.1. Definition of MSC

MSCs can be isolated from many tissues, including bone marrow, muscle, skin, and adipose tissue, and are characterized by the potential to differentiate into any tissue of mesenchymal origin, including muscle, fibroblasts, bone, tendon, ligament, and adipose tissue (1). However, the most common source is bone marrow. Currently, a precise definition of MSCs is elusive; thus, MSCs are usually defined functionally, rather than by the presence of specific surface markers. They adhere to cell culture dishes and do not express the surface markers that characterizes hematopoietic stem cells (2). Although precise phenotypic characterization varies in different studies, it is generally accepted that MSCs are negative for CD11b, CD14, CD31, CD34, and CD45 but are positive for CD29, CD44, CD73, CD105, CD106, and CD166 (3–7). They are uncommon in bone marrow, making up 0.001–0.01% of the total nucleated cells (8), and have a spindle-shaped fibroblast-like appearance in culture (see Fig. 1). They can be expanded dramatically in culture, maintaining their growth and multilineage potential (5). These properties of adherence in culture and multilineage potential are the most commonly accepted definitions of MSC.

Different subpopulations of MSCs have been described. The smallest and most rapidly dividing MSCs have been termed RS cells (recycling stem cells). Like all MSCs, they do not express hematopoietic stem cell surface markers, but in contrast to other MSC populations, these RS cells do express the receptor for stem cell factor, c-kit (CD117) (9). RS cells thus appear to be a more primitive form of MSC.

Another MSC subpopulation is the so-called multipotent adult progenitor cells (MAPCs). Unlike other MSC populations, these MAPCs isolated from bone marrow are immortal in culture

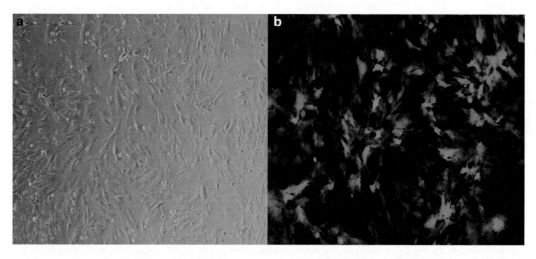

Fig. 1. Mesenchymal stem cells in culture. (**a**) MSCs in culture have a characteristic spindle shape and a fibroblast-like appearance. They are adherent to plastic. (**b**) MSCs transduced to express green fluorescent protein.

and can give rise to cell types from all three germ layers (10, 11). Finally, human bone marrow-derived multipotent stem cells (hBMSCs), like MAPCs, can give rise to cell types from all three germ layers and they share the ability of MSCs to engraft and differentiate to multiple lineages in a rodent model of postinfarction heart failure (12).

1.2. MSCs and the Immune System

One very important aspect of MSC function, both in health and disease, is the interaction of these cells with the immune system. MSCs form part of the hematopoietic stem cell niche in the bone marrow and are an integral part of maintaining homeostasis of these bone marrow stem cells. For an excellent review of MSCs and the immune system, please refer to Uccelli et al. (7). Two important qualities of MSC are that they are both immune-privileged and they are immunosuppressive. The former is a very attractive quality in cell-based therapies, because it may facilitate the use of allogeneic cells, rather than autologous cells, which offers many potential advantages. The latter property may also be helpful as it may suppress the immune response to the cell injections or to the damaged myocardium itself during myocardial infarction.

The immune-privileged status of MSCs is a vital aspect of their appeal as a potential source of cells for cardiac repair. An allogeneic cell therapy product would allow administration of an "off-the-shelf" product at the appropriate time following myocardial infarction. Currently, the requirement for autologous cell therapy demands a bone marrow aspirate or even cardiac biopsy at the time of myocardial infarction, which is uncomfortable and carries some degree of risk, as the patient is on antiplatelet and possibly anticoagulant therapy. In addition, these autologous cell sources limit the time when cells can be harvested, expanded, and given back to patients in clinically efficacious doses. An obvious benefit would thus be derived if one could deliver large numbers of preprepared cells that would not trigger an immune response.

How, then, do MSCs evade allo-rejection? MSCs express the major histocompatibility complex I (MHC I), but they have very low levels of MHC II. In addition, they lack the costimulatory molecules B7-1 (CD80), B7-2 (CD86), and CD40. With interferon gamma (IFNγ) stimulation, MSCs upregulate the expression of MHC II, a classic stimulator of the T-cell response, but the lack of the abovementioned costimulatory molecules prevents T-cell proliferation (13, 14). Thus, in the basal or stimulated state, MSCs evade allorecognition.

MSCs are immunosuppressive to activated T-lymphocytes. After activation of the T-cell receptor by specific antigens or mitogens, MSCs prevent T-cell proliferation without promoting apoptosis (13–15). Thus, MSCs not only evade the allogeneic T-cell response against themselves, but they also suppress that response to other allogeneic antigens that would usually incite a powerful

T-cell proliferative response. Aggarwal and Pittenger (16) examined the immunosuppressive role of MSCs by exposing MSCs to isolated subpopulations of immune cells, such as T_{Helper}-1 (T_H1), T_{Helper}-2 (T_H2), $T_{Regulatory}$ (T_R), and natural killer (NK) cells. They showed that MSCs interact with these cells and change their cytokine release profile from proinflammatory to anti-inflammatory. T_H1 and NK cells reduced their secretion of the proinflammatory cytokine IFNγ, while T_H2 cells increased their secretion of the immunoprotective interleukin-4. In addition, MSCs caused an increase in the number of T_R cells as well as the secretion of the anti-inflammatory cytokine interleukin-10. By these mechanisms, MSCs are not only able to evade rejection by the recipient immune system, but they are also able to modulate the recipient immune system to other stimuli. This makes it possible to deliver allogeneic cells, and perhaps to alter the immune/inflammatory response to myocardial infarction.

1.3. Preclinical Studies

Numerous animal studies have demonstrated the effectiveness of MSC therapy in repairing the damaged heart after myocardial infarction (12, 17–24). However, the exact mechanism(s) by which MSCs act remains incompletely described. It is likely that there are myriad effects from these versatile stem cells, rather than a single mechanism of action. Research has shown that a number of factors, both in vitro and in vivo, may contribute to the salutary effects of MSC therapy after MI.

1.3.1. MSC Survival and Engraftment

Before considering what becomes of the transplanted cells, and how they act, one must address the issue of cell survival. Only a small percentage of MSCs injected into the heart survive for prolonged periods of time (60). Magnetically labeled cells can be noninvasively tracked with magnetic resonance imaging (MRI). Using serial MRI imaging showed an incremental decline in the number of cells surviving to 8 weeks after direct endomyocardial injection in a porcine model of MI (17). Others have confirmed this phenomenon by examining cardiac tissue for surviving cells in animal models as well (24–26). These studies suggest that MSC numbers decline rapidly after injection into the myocardium and are barely detectable within several weeks. Importantly, the survival of implanted cells may be an important factor in determining how effectively they repair the damaged heart. Mangi and colleagues overexpressed the prosurvival gene Akt in MSCs and injected them into infarcted rat myocardium. They showed a dose (cell number)-dependent improvement in the regenerative capacity of these cells, which was far better than control MSCs (27). It is not known whether the increased survival per se was responsible for the improvement in function, or whether some other result of Akt upregulation led to the improved outcome.

But one can imagine that improved survival of implanted MSCs may translate into enhanced functional outcome.

The fate of therapeutically delivered MSCs remains controversial. Do they engraft and couple with the host myocardium electrically and mechanically? MSCs have demonstrated the ability to express the gap junction protein connexin 43 in vitro (28) and are capable of coupling with other MSCs and cardiomyocytes via connexin 43 (29). In vivo, labeled MCSs have been shown to couple to host cells via connexin 43 (21, 30, 31). These studies suggest that MSCs electrically couple to host cardiomyocytes, and do so better than other cell types, such as skeletal myoblasts. In addition, the numerous studies showing improvement in left ventricular function suggest that these cells also mechanically couple with the host myocardium.

1.3.2. Mechanisms of Action of MSCs for Cardiac Repair

It is clear that MSC therapy results in improved left ventricular function in preclinical studies. Although the mechanism is not known, a number of possible mechanisms exist (see Fig. 2) and these may act alone or in concert to achieve functional cardiac improvement. Do MSCs differentiate into new cardiomyocytes or other cell types? Do they enhance the neovascularization response seen after myocardial infarction? Do they act as reservoirs for paracrine mediators that are the actual effectors of the observed cardiac repair? Do they stimulate endogenous cardiac precursors to repair the damaged tissue?

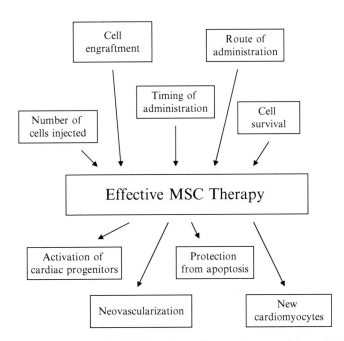

Fig. 2. Important aspects of MSC therapy in cardiac repair and possible mechanisms of action.

Differentiation of any type of bone marrow-derived stem cell into other cell types remains controversial. Differentiation of MSCs into cardiomyocytes has been described in vitro (32, 33). However, other groups show adoption of some cardiomyocyte protein expression by MSCs, but no true phenotype change into cardiomyocytes (28). In vivo, it has been shown that labeled MSCs injected into damaged hearts begin to express cardiomyocyte proteins (19, 34, 35, 60), suggestive, but not confirmatory, of differentiation. There is, however, evidence of cardiac regeneration. Serial multimodality imaging studies in a porcine model of myocardial infarction have been very illuminating in terms of assessing the mechanism or myocardial repair. Using serial scanning with multidetector computed tomography (MDCT), Amado and colleagues demonstrated that endomyocardial injection of MSCs following MI led to a progressive increase in thickness of the subendocardial rim of muscle tissue in the infarcted zone (18). This tissue developed contractile function and was histologically composed of small cardiomyocytes, excluding hypertrophy as the cause of the increased muscle mass in this region. The newly regenerated muscle was not of donor origin. Whether this new muscle arose from proliferation of adult cardiomyocytes or differentiation from cardiac stem cells is not known.

Adult cardiomyocytes have little capacity for self-renewal. They have widely been considered terminally differentiated cells with no ability to proliferate, but recent evidence has challenged this dogma. Endogenous cardiomyocytes at the edge of the infarct have been shown to express markers of cell cycle activation and mitotic figures have even been demonstrated (36). After MSC therapy in animal models, more cardiomyocytes express markers of cell cycle activation compared to control animals (17). Whether this truly reflects proliferation of these cardiomyocytes, or just expression of particular proteins remains unknown. Following MI, the cardiomyocytes at the infarct border region continue to undergo apoptosis throughout the remodeling process. After MSC therapy, the number of apoptotic cells is reduced (22, 23). This is likely to significantly contribute to the prevention of adverse LV remodeling seen with MSC therapy. Although the notion of cardiomyocyte proliferation remains hotly debated, it appears that MSC therapy tips the balance from prodeath, with ongoing cardiomyocyte loss and LV remodeling, to prosurvival, with less apoptosis and remodeling and possibly cardiomyocyte proliferation.

MSC therapy has several effects on cardiac vasculature. Schuleri and colleagues demonstrated that early after intramyocardial delivery of MSCs in a porcine model of MI, there is a dramatic but transient improvement in cardiac perfusion, demonstrated by cardiac MRI (23). This effect was seen in the first week after cell therapy, and not in the control animals, and this preceded the improvement in LV function. The increased perfusion was no longer seen 4 weeks later, at which time the improvement in left ventricular function was starting to be seen in the treated animals. It is unlikely that this early effect was due to the growth of new blood vessels; it was probably due to the release of vasoactive peptides from the MSCs acting in a paracrine manner to cause vasodilatation. After 8 weeks, the MSC-treated animals in this study showed a similar number of vessels to the control animals, but the MSC-treated animals had larger, more mature smooth muscle-coated vessels, rather than smaller capillaries. Although not all labs have demonstrated increased vascularity after MSC therapy for MI (22), most have demonstrated this phenomenon (20, 21, 24, 35, 60, 61). Thus, there is general agreement that there is an increase in vascularity after MSC therapy, and possibly an early paracrine-mediated increase in perfusion, but whether this is causally related to the improved cardiac function, or even a direct effect from the MSCs, remains unknown.

Recently a pool of resident cardiac stem cells has been described. They have been described by their surface markers or by their properties in cell culture, but there is now general agreement of the existence of resident cardiac stem cells. In vitro, MSCs induce chemotaxis, improve survival, and promote cardiac differentiation of cardiac progenitor cells (CPCs) (37). It is possible that at least part of the functional improvement in LV contraction due to MSC therapy is due to their effects on the resident CPCs, however, this remains to conclusively shown.

The mechanism of action of MSCs (and indeed all stem cell therapy) in cardiac repair remains partially understood and highly controversial. Multiple processes may be acting in parallel, making this a very complex area of study. Further research into the precise mechanism may facilitate improvement of the technique to achieve even more functional repair for damaged hearts, which will serve to benefit patients in the long run.

1.3.3. Routes of Administration

There are a number of possible routes of administration of stem cells, and the benefits of one over another have yet to be established. MSCs have been used for cardiac repair in preclinical trials via numerous delivery routes, including intravenous (IV) (38–40), intracoronary artery (IC) (41, 42), direct epicardial injection through a thoracotomy (24, 43), and endomyocardial injection via catheter-based delivery systems (17, 44, 45). The least invasive method to deliver the cells is IV; however, this results in more cells being filtered by the lungs and fewer reaching and engrafting in the heart (44). The next least invasive method is direct injection of cells into the coronary artery, which can readily be achieved in any cardiac catheterization laboratory. However, like IV delivery, this method of intravascular injection also leads to low rates of cell retention in the heart (45). Methods to increase the delivery of the cells outside the vascular space and into the tissue that needs repair have been developed. During thoracotomy for bypass surgery or other reasons, epicardial injection of cells into the myocardium under direct visualization is readily achieved. This technique allows the operator to visualize the exact area of the heart that requires cell therapy and precise multiple injections can be delivered accurately. The downside to this technique is that thoracotomy is invasive, costly, requires prolonged hospitalization, and has its own morbidity associated with it. Thus, catheter-based less invasive strategies have been devised. Several catheter injection systems have been devised to facilitate cell injection into the myocardium from the endocardial surface of the left ventricular myocardium. These can be guided in real-time using fluoroscopy or electromechanical mapping and have been demonstrated to be safe and effective in animal models and have thus advanced to clinical trials in patients.

Few studies have directly compared the different modalities of MSC delivery. Freyman and colleagues (44) compared IV, IC, and endocardial injection of MSCs in a porcine model of MI. They showed that IV injection of MSCs resulted in lower rates of cell engraftment in the infarct zone than IC or endocardial injection. Furthermore, both IV and IC had higher rates of distal organ cell engraftment than endocardial injection. Perin et al. (45) compared IC and endocardial injection routes in a canine model of MI. They demonstrated that endocardial injection results in higher cell retention rates and better functional improvement than IC injection. However, these studies differed in the number of cells injected, the method of IC injection and the animal species used. Similar comparisons are needed, in conjunction with dose finding studies to define the optimal number of cells that need to be delivered by each route, before this can become a standard therapy for patients.

1.4. Human Clinical Trials

The data in the literature about the effects of MSC therapy for ischemic heart failure is emerging. Whereas bone marrow cells have been used by many investigators and injected IC after MI, the field has progressed more slowly with MSC therapy. There are numerous preclinical studies investigating the appropriate timing and number of cells to inject, which will inform clinical trial investigators to design appropriate studies. Some small pilot studies have been reported which demonstrate safety and efficacy of using this cell type in humans, and now is the appropriate time to advance to phase 3 clinical trials to definitively assess the efficacy of this therapy.

Hare and colleagues performed a double-blind, randomized, placebo-controlled, dose escalation study of intravenous allogeneic MSCs in 53 patients following MI. They showed safety of the MSC therapy, with improvement in ventricular arrhythmias and pulmonary function in the cell therapy groups. In addition, they showed improved LV function in the subset of patients who had anterior MI (62).

Mohyeddin and colleagues (46) performed a pilot study of eight patients with old MI undergoing revascularization either by percutaneous coronary intervention (PCI) or coronary artery bypass graft (CABG) surgery, and injected autologous MSCs to test safety and feasibility. The PCI patients received IC infusion of MSCs and the CABG patients received direct epicardial injection of MSCs. They demonstrated smaller perfusion defect, improved LV ejection fraction, and improved heart failure functional class. Chen and colleagues (47) studied 69 patients after PCI for acute MI. They randomized them to IC administration of autologous MSCs or normal saline approximately 3 weeks post-MI. Follow-up at 3 months demonstrated a smaller perfusion defect in the treated group, assessed by positron emission tomography (PET), and improved LV ejection fraction that was sustained to 6 months of follow-up. The ejection fraction improved from $47 \pm 9\%$ before treatment to $67 \pm 3\%$ after 6 months in the MSC-treated group and from $48 \pm 10\%$ to $54 \pm 5\%$ in the control group ($p=0.01$). These preliminary studies are encouraging and suggest that autologous MSC therapy is safe and feasible. However, larger randomized controlled trails are needed to confirm safety and demonstrate efficacy.

With a wealth of preclinical studies guiding the way, MSC therapy for human cardiac repair is entering clinical trials. Preliminary study results are very encouraging. The results of larger randomized controlled trials are eagerly anticipated. If the human clinical trials mirror the preclinical data, MSC therapy may become a formidable weapon in the fight against heart failure. This could substantially improve patient care and reduce health care costs.

1.5. Cell Tracking and Evaluation of MSC Therapy

A number of methods exist to track the survival, engraftment, distribution, and reparative efficacy of MSCs in preclinical models, in both small and large animals. All of these methods have limitations and advantages compared to the other methods, and none stands out as a single best option for cell tracking. The currently utilized methods for cell tracking are outlined in Table 1. Some methods require examination of harvested cardiac tissue, and therefore their use is limited to animal models only. Other methods can noninvasively track cells and thus can potentially also be used in clinical applications of MSC therapy.

1.5.1. Tissue Assessment of MSC Therapy

Several methods exist for tracking cells by examining harvested tissue. The finding of Y-chromosome positive cells suggests male origin of the cells, and some investigators have used this method to find male donor cells in female hearts (where none should exist normally). Quaini and colleagues used this method in the hearts of male recipients of female-donor cardiac transplants, and showed that some Y chromosome positive cells of host origin invade the recipient heart (48). Others have used this method of fluorescence in-situ hybridization (FISH) following MSC therapy to follow the fate of implanted cells (44, 60). This method of cell identification

Table 1
Methods of labeling and tracking MSCs in tissue and noninvasively

Target	Cellular location	Noninvasively detectable
Membrane label e.g., DiI	Cell surface	No
DNA label e.g., DAPI	Nuclear	No
Transfection with reporter genes e.g., β-gal, GFP	Cytoplasmic	No
In-situ hybridization for Y chromosome in female recipient of male cells	Nuclear	No
PCR for transgene or Y chromosome	Nuclear/cytoplasmic	No
Iron labeling for MRI	N/A	Yes
Iridium labeling for neutron activation analysis (NAA)	N/A	Yes
Indium labeling for SPECT	N/A	Yes
PET imaging of transfected cells carrying an emitting reporter gene	N/A	Yes

DiI 1,1″-dioleyl-3,3,3″,3″-tetramethylindocarbocyanine methanesulphonate, *DNA* deoxyribonucleic acid, *DAPI* 4′,6-diamino-2-phenylindole, *β-gal* β-galactosidase, *GFP* green fluorescent protein, *PCR* polymerase chain reaction, *MRI* magnetic resonance imaging, *SPECT* single photon emission computed tomography, *PET* positron emission tomography

has some clear advantages. It is the only method that does not require labeling of the cells in advance, and therefore does not interfere with cellular functioning, yet allows for long-term labeling with 100% efficiency. However, sections must include the nucleus to demonstrate the Y chromosome, and therefore this may underestimate the number of Y chromosome positive cells in the section. Furthermore, the tissue preparation required for in-situ hybridization makes costaining to further identify the cells quite difficult. Another method to track the existence of implanted make cells is to perform polymerase chain reaction (PCR) for the presence of Y-chromosome (26, 49). This is technically easier than FISH and is more sensitive in detecting the presence of Y-chromosome, but can give no hint as to the fate of the implanted cells. FISH, on the other hand, can identify where and what type of cell has a Y-chromosome, and this gives some clue to the fate of the cell.

Membrane-bound cell labels, such as DiI (1,1″-dioleyl-3,3,3″,3″-tetramethylindocarbocyanine methanesulphonate), are easily used, reliably label a cell for some time, and can easily be seen in tissue sections. However, they can be lost from the surface of the cells and may be taken up by the labeled cells or by other cells, reducing the specificity of this technique. In addition, the membrane labels become less distinct with time, thus limiting the sensitivity as well. Gene transfection with reporter genes is a well-accepted method of reliably labeling transplanted cells. However, the very process of transfection may alter the cellular function. This is of limited benefit in predicting the response in human trials, where transfected cells are less likely to be used, because of the possibility of unanticipated adverse events.

1.5.2. Noninvasive Imaging of MSC Therapy

The ability to track injected cells noninvasively is attractive because it can be used for serial follow-up, whereas examination of harvested tissue examines just one time-point per subject. In addition, unlike tissue harvesting, it has the potential for use in clinical trials. Furthermore, serial noninvasive imaging has the potential to demonstrate time-dependent changes in cardiac structure and function may help elucidate the mechanisms of cardiac repair (18). Unfortunately, no perfect imaging for transplanted cells exists at this time and all currently available methods have their unique strengths and weaknesses (50). The ideal imaging agent for stem cell therapy should be safe and nontoxic, biocompatible and highly specific for the target labeled cells, with no leak to surrounding cells. In addition, the ideal imaging system should be sensitive enough to see an increase in activity with cell proliferation and a decrease with cell death (50). Currently, no imaging modality has all these properties.

Magnetic particle labeling of MSCs makes them visible on MRI and has no effect on the viability or proliferative capacity of the cells (51). This seems a very attractive way to follow cells, particularly as MRI can also be used as an accurate method to measure global and regional left ventricular function and infarct size; therefore, a number of authors have used this method (17, 52). However, the persistence of iron oxide detected using MRI, as used in these studies, has since been shown to overestimate the number of surviving cells (25, 53) and is predominantly found engulfed by cardiac macrophages. Thus, enthusiasm for this method of cell labeling has been dampened recently.

Radionuclide imaging has progressed significantly in recent years and has high intrinsic sensitivity compared to MRI. Improvements in spatial resolution (1–2 mM) have made this modality particularly suited to cell tracking (50). In addition, clinically available radioisotopes, such as 99 m-Technetium, have been used in preclinical models to label and successfully track MSC distribution (54), and therefore the translation to human clinical trials will be easier. Indium-111 has been shown to effectively label MSCs without any untoward effects on viability or differentiation capacity (55). The safety of this radiotracer was called into question by other authors, who showed that Indium-111 labeled CD34+ hematopoietic stem cells have impaired proliferative and differentiation capacity (56). Further studies to determine the safety profile of each radiotracer for MSCs are needed before the optimal tracer for clinical trials can be decided.

**1.6. Additional
Considerations**

*1.6.1. Genetic Modification
and MSCs*

In addition to their own therapeutic effects, MSCs can be genetically modified to enhance their effectiveness. Mangi et al. modified MSCs to over express the prosurvival factor Akt (27). Rendering the MSCs resistant to hypoxic cell death, they were able to enhance the efficacy of myocardial repair. In fact, they achieved almost complete regeneration of the infarcted myocardium with no pathological tissue growth. Their results may be due to enhancing the survival of MSCs, allowing more to survive and deliver their own therapeutic effects; or it may have been due to the overexpression of Akt having some other effect(s). Here the MSCs may have been the effectors of repair, or they may simply have been a vehicle for the effector of repair. In another example of genetic modification of MSCs, overexpression of angiogenin in MSCs resulted in resistance to hypoxic cell death (57). When transplanted into a rat model of MI, the angiogenin-overexpressing MSCs achieved a greater reduction in remodeling parameters and better outcomes than standard MSCs. Thus, producing MSCs that overexpress various factors can be used as a valuable research tool. Viral transfection techniques have an uphill battle to reach clinical application, but will continue to serve as a valuable research tool. Novel nonviral transduction methodologies are eagerly anticipated.

1.6.2. Advantages of MSCs

MSCs have some clear potential advantages over other stem cell types. They are relatively easy to culture to large numbers and are stable through many passages in vitro. Thus they can easily be scaled up to clinically relevant doses. Combined with their immune privileged status, this means they can potentially be given as an allogeneic product "off the shelf." This would be therapeutically advantageous over other cell types because they could be administered at any time, even during reperfusion of acute myocardial infarction. Although the optimal timing of cell therapy remains unknown, having the luxury of a premade product ready for delivery anytime is certainly an advantage. Furthermore, the ability to deliver the product IV allows it to be given anywhere, without the need for cardiac catheterization laboratories, expensive mapping and endocardial injection systems, or even open thoracotomy.

*1.6.3. Potential Limitations
of MSC Therapy*

Some safety concerns have been raised regarding MSC therapy. Vulliet and colleagues performed IC injection of MSCs in dogs and found that they caused microinfarctions (58). Freyman and colleagues administered IC MSCs in a porcine model of MI and observed slow flow in the coronary arteries in half their subjects (44). MSCs are larger than other stem cell types and are characterized by their adherence to plastic in vitro. It is not surprising, therefore, that some microvascular obstruction occurs with IC delivery. However, in the limited clinical experience described above, this does not appear to be a problem.

Both MSCs and bone marrow-derived stem cells have recently been linked to calcification and possibly ossification of the heart in a murine model of MI (59). These stem cell types are precursors for bone formation and it is conceivable that, under the appropriate environmental conditions, they are capable of contributing to ectopic calcification or even bone formation. This has not been reported by multiple other studies in many species. Ongoing safety studies should address this issue in all species.

1.7. Future Directions

A wealth of preclinical experience has accumulated about the safety and efficacy of mesenchymal stem cell therapy for cardiac repair. Although the precise mechanism of action of these cells remains incompletely described, there has been clear demonstration of myocardial regeneration and neovascularization. Future studies will address the efficacy of MSC therapy in adult and aged animals, rather than young adult animals, as well as in combination with standard post-MI therapies, such as angiotensin converting enzyme inhibitors and beta-blockers.

There is a need for dose escalation studies delivering various numbers of cells via the different possible routes to optimize therapy before MSCs can be realized as a potential clinical treatment. Furthermore, most studies have been limited to postinfarction heart failure, with relatively few studies performed in nonischemic heart failure models, so the benefits of MSC therapy in this setting is less clear and further studies are needed.

MSC therapy promises to repair damaged myocardium for thousands of heart failure patients. Realizing this promise will require ongoing laboratory and preclinical studies, perfored in parallel with rigorously controlled clinical trials.

2. Materials

2.1. Isolation of Bone Marrow Mononuclear Cells by Density Gradient Centrifugation

2.1.1. Equipment

1. Microscope.
2. Biological Safety Cabinet (BSC).
3. Centrifuge.
4. Vacuum Pump.
5. Coulter A^cT Diff™ Analyzer.

2.1.2. Supplies

1. 4×4 gauze, sterile.
2. Conical tubes, 50 ml and 250 ml, sterile.
3. Pipettes, 1 ml, 10 ml, 25 ml, 50 ml and aspirating pipettes, sterile.
4. Pipette-aid.

5. Surgical gowns, shoe covers, head cover, and mask, all sterile.

6. Surgical gloves, sterile.

7. Vacuum collection flask with two associated sterile tubing sets, sterile.

8. Pipette tips, sterile.

9. Microscope slides for cytospin.

2.1.3. Reagents

1. Lymphocyte Separation Medium (LSM) (Lonza).

2. Plasma-Lyte A, 1× (Baxter).

3. Trypan Blue.

4. Human Serum Albumin (HSA), 25%.

2.1.4. Wash Medium

1. Plasma-Lyte A, 1×.

2. 25% HSA, 1% final concentration.

2.2. Culture and Expansion of Human Marrow Mesenchymal Stem Cells

2.2.1. Equipment

1. Microscope.

2. CO_2 Incubators.

3. Biological Safety Cabinet (BSC).

4. Centrifuge.

5. Vacuum Pump.

6. Hemocytometer.

7. Coulter A^{c}T Diff™ Analyzer.

8. Water bath, set a 37°C (to thaw reagents required to prepare the necessary media).

9. Pipette-aid.

2.2.2. Supplies

1. Gauze 4×4, sterile.

2. Conical centrifuge tubes, 50 ml and 250 ml, sterile.

3. Pipettes, 1 ml, 10 ml, 25 ml, 50 ml and aspirating pipettes, all sterile.

4. Surgical gowns, shoe covers, head cover, and mask, all sterile.

5. Surgical gloves, sterile.

6. Vacuum collection flask with two associated sterile tubing sets, sterile.

7. Pipette tips, 200 µl, sterile.

8. Culture flasks, Nunc ΔT185 Cell Culture Treated, sterile.

9. Cell Scraper, sterile.

2.2.3. Reagents

1. Alpha Minimal Essential Medium (MEM), 1×.

2. Penicillin–Streptomycin–Glutamine (100×), liquid.

3. Fetal Bovine Serum (FBS), Gamma Irradiated.

4. Trypsin EDTA, 1×; 0.12% Gamma Irradiated.

5. Phosphate Buffered Saline (DPBS), Ca^{2+} and Mg^{2+} Free.

6. l-Glutamine 200 mM (100×).

7. Human Serum Albumin (HSA), 25%.

8. Trypan blue.

2.2.4. Media

1. Wash Medium.
   (a) DPBS (Ca^{2+} and Mg^{2+} free).
   (b) 25% Human Serum Albumin, final concentration 1%.
2. Complete Culture Medium.
   (a) Alpha MEM medium.
   (b) Penicillin–Streptomycin–Glutamine (100×), liquid.
   (c) FBS, final concentration 20%.

3. Methods

3.1. Isolation of Bone Marrow Mononuclear Cells by Density Gradient Centrifugation

3.1.1. Product Processing

1. Dilute the BM product 1:1 with wash medium.

2. Aliquot 15 ml of LSM into 50 ml tubes (the number of tubes needed will be the total volume of the diluted BM product divided by 30).

3. Overlay 30 ml of the diluted bone marrow aspirate on top of LSM (30 ml of diluted bone marrow aspirate over 15 ml of LSM).

4. Carefully, transfer the cells to a centrifuge and spin at $800 \times g$ for 30 min, at room temperature, with the centrifuge brake "OFF" (see Note 1).

5. After centrifugation, using a sterile transfer pipette, collect the interface from each 50 ml conical tube and transfer it into clean 50 ml conicals.

6. Bring the volume in each tube up to 50 ml with Wash Medium.

7. Centrifuge at $500 \times g$ for 10 min, with brake set to "Low."

8. Using an aspirating pipette, remove almost all supernatant, leaving 1–2 ml in each tube.

9. Gently resuspend the cell pellet in each conical. Combine cells from all the conical tubes into a single clean 50 ml conical. Bring the total volume up to 50 ml with wash medium.

10. Centrifuge at $500 \times g$, for 10 min, at room temperature, with the centrifuge brake set to "Low."

11. Using an aspirating pipette, remove almost all supernatant, leaving 1–2 ml of the supernatant in the conical. Gently resuspend the cell pellet and bring the volume up to 50 ml using fresh Wash Medium.

12. Perform a cell count and viability using trypan blue.

3.2. Culture and Expansion of Human Marrow Mesenchymal Stem Cells

Prepare the medium as per Subheading 2.2.4

3.2.1. Medium Preparation

3.2.2. Cell Culture and Expansion (P0)

1. Once the MNCs are prepared, label ten flasks.

2. Seed each T-185 flask by dividing the mononuclear cells equally among the ten flasks and bring to a final volume of 25 ml with Complete Culture Medium.

3. Incubate in tissue culture incubator, with 5% CO_2, for at least 72 h, to allow the cells to adhere to the flask.

4. After 72 h from the initial plating, remove the flask from the incubator and remove the unattached cells by tilting the flask while holding it upright and aspirating off all of the media from the bottom corner of the flask.

5. Add 25 ml of Complete Culture Medium.

6. Place the flask in a tissue culture incubator, at 37°C and 5% CO_2.

3.2.3. Feeding the Cells

1. Cells should be fed every 3–4 days until harvest, when attached cells are confluent.

2. Remove the flasks from the incubator. Observe under the microscope to determine the extent to which the cells are confluent. The cells will be split when they are >80% confluent. In addition, examine each tissue culture flask for absence of contamination. If contamination is present, discard the flask or the whole preparation.

3. Using vacuum aspiration, remove the culture media from the T-180 flasks.

4. Add fresh 25 ml of Complete Culture Medium.

5. Place back into the 37°C incubator with 5% CO_2, for further culture.

3.2.4. Harvesting Cells Using Trypsin EDTA

1. When the cells are approximately >80% confluent (day 14 of culture), observe each tissue culture flask under the microscope to examine the cells for absence of contamination. If contamination is present, discard the flask or the whole preparation, if it becomes necessary.

2. Remove the medium from each flask using vacuum aspiration.

3. Add 25 ml of Wash Medium and swirl it around the flask.

4. Aspirate off the Wash Medium and add 10 ml of Trypsin EDTA to each flask. Return the flask to the 37°C CO_2 incubator for not more than 8 min. Make sure to monitor the

cells at 4 min intervals under the microscope, as leaving the Trypsin in the flask for a prolonged period of time may be damaging to the cells.

5. If there are still attached cells, use a sterile cell scraper to gently lift the adhered cells by moving the scraper across the bottom of each flask.

6. Neutralize Trypsin activity by adding 15 ml of Complete Culture Medium to each flask. Swirl the medium around the flask to make sure all the cells are in suspension.

7. Transfer the detached cells from each flask to a sterile 50 ml conical, using a 25 ml pipette.

8. Add 25 ml of Wash Media to the flask and swirl to remove any remaining cells. Using a 25 ml pipette, collect the Wash Media from the flask and add it to the 50 ml conical with the detached cells.

9. Centrifuge the 50 ml conicals at $500 \times g$ for 10 min, at room temperature, with brake set to "Low."

10. Using a vacuum system, aspirate the supernatant to leave 1–2 ml of medium and resuspend the pellet in the 50 ml conical. Bring the volume up to 40 ml with Complete Culture Medium.

11. Take a 100 µl sample and perform a cell count and viability using trypan blue (see Note 2).

4. Notes

1. The culture of bone marrow cells (BM) to isolate and expand mesenchymal stem cells (MSCs) is inhibited by red blood cells (RBCs) and granulocytes. In particular, the large number of RBCs cover the surface of tissue culture flasks. Isolation of the mononuclear cells (MNC) by density centrifugation, removes the RBCs and granulocytes, providing an optimal starting cell population for isolation and expansion of MSC.

2. After the above steps, MSCs may be cryopreserved or expansion continued in new flasks.

References

1. Caplan, A. I. (1991) Mesenchymal stem cells. *J Orthop Res* **9**, 641.
2. Schuleri, K. H., Boyle, A. J., and Hare, J. M. (2007) Mesenchymal stem cells for cardiac regenerative therapy. *Handb Exp Pharmacol* **180**, 195–218.
3. Ryan, J. M., Barry, F. P., Murphy, J. M., and Mahon, B. P. (2005) Mesenchymal stem cells avoid allogeneic rejection. *J Inflamm* **2**, 8.
4. Kemp, K. C., Hows, J., and Donaldson, C. (2005) Bone marrow-derived mesenchymal stem cells. *Leuk Lymphoma* **46**, 1531–44.

5. Pittenger, M. F., and Martin, B. J. (2004) Mesenchymal stem cells and their potential as cardiac therapeutics. *Circ Res* **95**, 9–20.

6. Zimmet, J., and Hare, J. (2005) Emerging role for bone marrow derived mesenchymal stem cells in myocardial regenerative therapy. *Basic Res Cardiol* **100**, 471.

7. Uccelli, A., Moretta, L., and Pistoia, V. (2008) Mesenchymal stem cells in health and disease. *Nat Rev Immunol* **8**, 726–36.

8. Pittenger, M. F., Mackay, A. M., Beck, S. C., Jaiswal, R. K., Douglas, R., Mosca, J. D. et al. (1999) Multilineage potential of adult human mesenchymal stem cells. *Science* **284**, 143–7.

9. Prockop, D. J., Sekiya, I., and Colter, D. C. (2001) Isolation and characterization of rapidly self-renewing stem cells from cultures of human marrow stromal cells. *Cytotherapy* **3**, 393–6.

10. Jiang, Y., Jahagirdar, B. N., Reinhardt, R. L., Schwartz, R. E., Keene, C. D., Ortiz-Gonzalez, X. R., et al. (2002) Pluripotency of mesenchymal stem cells derived from adult marrow. *Nature* **418**, 41–9.

11. Jiang, Y., Vaessen, B., Lenvik, T., Blackstad, M., Reyes, M., and Verfaillie, C. M. (2002) Multipotent progenitor cells can be isolated from postnatal murine bone marrow, muscle, and brain. *Exp Hematol* **30**, 896–904.

12. Yoon, Y. S., Wecker, A., Heyd, L., Park, J.-S., Tkebuchava, T., Kusano, K., et al. (2005) Clonally expanded novel multipotent stem cells from human bone marrow regenerate myocardium after myocardial infarction. *J Clin Invest* **115**, 326–38.

13. Bartholomew, A., Sturgeon, C., Siatskas, M., Ferrer, K., McIntosh, K., Patil, S., et al. (2002) Mesenchymal stem cells suppress lymphocyte proliferation in vitro and prolong skin graft survival in vivo. *Exp Hematol* **30**, 42–8.

14. Tse, W. T., Pendleton, J. D., Beyer, W. M., Egalka, M. C., and Guinan, E. C. (2003) Suppression of allogeneic T-Cell proliferation by human marrow stromal cells: implications in transplantation. *Transplantation* **75**, 389–97.

15. Di Nicola, M., Carlo-Stella, C., Magni, M., Milanesi, M., Longoni, P. D., Matteucci, P., et al. (2002) Human bone marrow stromal cells suppress T-lymphocyte proliferation induced by cellular or nonspecific mitogenic stimuli. *Blood* **99**, 3838–43.

16. Aggarwal, S., and Pittenger, M. F. (2005) Human mesenchymal stem cells modulate allogeneic immune cell responses. *Blood* **105**, 1815–22.

17. Amado, L. C., Saliaris, A. P., Schuleri, K. H., St. John, M., Xie, J.-S., Cattaneo, S., et al. (2005) Cardiac repair with intramyocardial injection of allogeneic mesenchymal stem cells after myocardial infarction. *Proc Natl Acad Sci U S A* **102**, 11474–9.

18. Amado, L. C., Schuleri, K. H., Saliaris, A. P., Boyle, A. J., Helm, R., Oskouei, B., et al. (2006) Multimodality noninvasive imaging demonstrates in vivo cardiac regeneration after mesenchymal stem cell therapy. *J Am Coll Cardiol* **48**, 2116–24.

19. Shake, J. G., Gruber, P. J., Baumgartner, W. A., Senechal, G., Meyers, J., Redmond, J. M., et al. (2002) Mesenchymal stem cell implantation in a swine myocardial infarct model: engraftment and functional effects. *Ann Thorac Surg* **73**, 1919–26.

20. Olivares, E. L., Ribeiro, V. P., Werneck de Castro, J. P. S., Ribeiro, K. C., Mattos, E. C., Goldenberg, R. C. S., et al. (2004) Bone marrow stromal cells improve cardiac performance in healed infarcted rat hearts. *Am J Physiol Heart Circ Physiol* **287**, H464–70.

21. Nagaya, N., Fujii, T., Iwase, T., Ohgushi, H., Itoh, T., Uematsu, M., et al. (2004) Intravenous administration of mesenchymal stem cells improves cardiac function in rats with acute myocardial infarction through angiogenesis and myogenesis. *Am J Physiol Heart Circ Physiol* **287**, H2670–6.

22. Berry, M. F., Engler, A. J., Woo, Y. J., Pirolli, T. J., Bish, L. T., Jayasankar, V., et al. (2006) Mesenchymal stem cell injection after myocardial infarction improves myocardial compliance. *Am J Physiol Heart Circ Physiol* **290**, H2196–203.

23. Schuleri, K. H., Amado, L. C., Boyle, A. J., Centola, M., Saliaris, A. P., Gutman, M. R., et al. (2008) Early improvement in cardiac tissue perfusion due to mesenchymal stem cells. *Am J Physiol Heart Circ Physiol* **294**, H2002–11.

24. Imanishi, Y., Saito, A., Komoda, H., Kitagawa-Sakakida, S., Miyagawa, S., Kondoh, H., et al. (2008) Allogenic mesenchymal stem cell transplantation has a therapeutic effect in acute myocardial infarction in rats. *J Mol Cell Cardiol* **44**, 662–71.

25. Terrovitis, J., Stuber, M., Youssef, A., Preece, S., Leppo, M., Kizana, E., et al. (2008) Magnetic resonance imaging overestimates ferumoxide-labeled stem cell survival after transplantation in the heart. *Circulation* **117**, 1555–62.

26. de Macedo Braga, L., Lacchini, S., Schaan, B., Rodrigues, B., Rosa, K., De Angelis, K., et al. (2008) In situ delivery of bone marrow cells and mesenchymal stem cells improves cardiovascular function in hypertensive rats submitted

to myocardial infarction. *J Biomed Sci* **15**, 365–74.

27. Mangi, A. A., Noiseux, N., Kong, D., He, H., Rezvani, M., Ingwall, J. S., et al. (2003) Mesenchymal stem cells modified with Akt prevent remodeling and restore performance of infarcted hearts. *Nat Med* **9**, 1195–201.

28. Rose, R. A., Jiang, H., Wang, X., Helke, S., Tsoporis, J. N., Gong, N., et al. (2008) Bone marrow-derived mesenchymal stromal cells express cardiac-specific markers, retain the stromal phenotype and do not become functional cardiomyocytes in vitro. *Stem Cells* **26**, 2884–92.

29. Valiunas, V., Doronin, S., Valiuniene, L., Potapova, I., Zuckerman, J., Walcott, B., et al. (2004) Human mesenchymal stem cells make cardiac connexins and form functional gap junctions. *J Physiol (Lond)* **555**, 617–26.

30. Nagaya, N., Kangawa, K., Itoh, T., Iwase, T., Murakami, S., Miyahara, Y., et al. (2005) Transplantation of mesenchymal stem cells improves cardiac function in a rat model of dilated cardiomyopathy. *Circulation* **112**, 1128–35.

31. Mills, W. R., Mal, N., Kiedrowski, M. J., Unger, R., Forudi, F., Popovic, Z. B., et al. (2007) Stem cell therapy enhances electrical viability in myocardial infarction. *J Mol Cell Cardiol* **42**, 304–14.

32. Makino, S., Fukuda, K., Miyoshi, S., Konishi, F., Kodama, H., Pan, J., et al. (1999) Cardiomyocytes can be generated from marrow stromal cells in vitro. *J Clin Invest* **103**, 697–705.

33. Pijnappels, D. A., Schalij, M. J., Ramkisoensing, A. A., Tuyn, J. V., Vries, A. A. F. D., Laarse, A. V. D., et al. (2008) Forced alignment of mesenchymal stem cells undergoing cardiomyogenic differentiation affects functional integration with cardiomyocyte cultures. *Circ Res* **103**, 167–76.

34. Toma, C., Pittenger, M. F., Cahill, K. S., Byrne, B. J., and Kessler, P. D. (2002) Human mesenchymal stem cells differentiate to a cardiomyocyte phenotype in the adult murine heart. *Circulation* **105**, 93–8.

35. Gojo, S., Gojo, N., Takeda, Y., Mori, T., Abe, H., Kyo, S., et al. (2003) In vivo cardiovasculogenesis by direct injection of isolated adult mesenchymal stem cells. *Exp Cell Res* **288**, 51–9.

36. Beltrami, A. P., Urbanek, K., Kajstura, J., Yan, S.-M., Finato, N., Bussani, R., et al. (2001) Evidence that human cardiac myocytes divide after myocardial infarction. *N Engl J Med* **344**, 1750–7.

37. Nakanishi, C., Yamagishi, M., Yamahara, K., Hagino, I., Mori, H., Sawa, Y., et al. (2008) Activation of cardiac progenitor cells through paracrine effects of mesenchymal stem cells. *Biochem Biophys Res Commun* **374**, 11–6.

38. Krause, U., Harter, C., Seckinger, A., Wolf, D., Reinhard, A., Bea, F., et al. (2007) Intravenous delivery of autologous mesenchymal stem cells limits infarct size and improves left ventricular function in the infarcted porcine heart. *Stem Cells Dev* **16**, 31–8.

39. Price, M. J., Chou, C.-C., Frantzen, M., Miyamoto, T., Kar, S., Lee, S., et al. (2006) Intravenous mesenchymal stem cell therapy early after reperfused acute myocardial infarction improves left ventricular function and alters electrophysiologic properties. *Int J Cardiol* **111**, 231–9.

40. Halkos, M., Zhao, Z.-Q., Kerendi, F., Wang, N.-P., Jiang, R., Schmarkey, L., et al. (2008) Intravenous infusion of mesenchymal stem cells enhances regional perfusion and improves ventricular function in a porcine model of myocardial infarction. *Basic Res Cardiol* **103**, 525–36.

41. Lim, S. Y., Kim, Y. S., Ahn, Y., Jeong, M. H., Hong, M. H., Joo, S. Y., et al. (2006) The effects of mesenchymal stem cells transduced with Akt in a porcine myocardial infarction model. *Cardiovasc Res* **70**, 530–42.

42. Molina, E. J., Palma, J., Gupta, D., Torres, D., Gaughan, J. P., Houser, S., et al. (2008) Improvement in hemodynamic performance, exercise capacity, inflammatory profile, and left ventricular reverse remodeling after intracoronary delivery of mesenchymal stem cells in an experimental model of pressure overload hypertrophy. *J Thorac Cardiovasc Surg* **135**, 292–99.e1.

43. Makkar, R. R., Price, M. J., Lill, M., Frantzen, M., Takizawa, K., Kleisli, T., et al. (2005) Intramyocardial injection of allogenic bone marrow-derived mesenchymal stem cells without immunosuppression preserves cardiac function in a porcine model of myocardial infarction. *J Cardiovasc Pharmacol Ther* **10**, 225–33.

44. Freyman, T., Polin, G., Osman, H., Crary, J., Lu, M., Cheng, L., et al. (2006) A quantitative, randomized study evaluating three methods of mesenchymal stem cell delivery following myocardial infarction. *Eur Heart J* **27**, 1114–22.

45. Perin, E. C., Silva, G. V., Assad, J. A. R., Vela, D., Buja, L. M., Sousa, A. L. S., et al. (2008) Comparison of intracoronary and transendocardial delivery of allogeneic mesenchymal

cells in a canine model of acute myocardial infarction. *J Mol Cell Cardiol* **44**, 486–95.

46. Mohyeddin-Bonab, M., Mohamad-Hassani, M. R., Alimoghaddam, K., Sanatkar, M., Gasemi, M., Mirkhani, H., et al. (2007) Autologous in vitro expanded mesenchymal stem cell therapy for human old myocardial infarction. *Arch Iran Med* **10**, 467–73.

47. Chen, S.-L., Fang, W.-W., Ye, F., Liu, Y.-H., Qian, J., Shan, S.-J., et al. (2004) Effect on left ventricular function of intracoronary transplantation of autologous bone marrow mesenchymal stem cell in patients with acute myocardial infarction. *Am J Cardiol* **94**, 92–5.

48. Quaini, F., Urbanek, K., Beltrami, A. P., Finato, N., Beltrami, C. A., Nadal-Ginard, B., et al. (2002) Chimerism of the transplanted heart. *N Engl J Med* **346**, 5–15.

49. Müller-Ehmsen, J., Krausgrill, B., Burst, V., Schenk, K., Neisen, U. C., Fries, J. W. U., et al. (2006) Effective engraftment but poor mid-term persistence of mononuclear and mesenchymal bone marrow cells in acute and chronic rat myocardial infarction. *J Mol Cell Cardiol* **41**, 876–84.

50. Zhang, S. J., and Wu, J. C. (2007) Comparison of imaging techniques for tracking cardiac stem cell therapy. *J Nucl Med* **48**, 1916–9.

51. Arbab, A. S., Yocum, G. T., Rad, A. M., Khakoo, A. Y., Fellowes, V., Read, E. J., et al. (2005) Labeling of cells with ferumoxides-protamine sulfate complexes does not inhibit function or differentiation capacity of hematopoietic or mesenchymal stem cells. *NMR Biomed* **18**, 553–9.

52. Kraitchman, D. L., Heldman, A. W., Atalar, E., Amado, L. C., Martin, B. J., Pittenger, M. F., et al. (2003) In vivo magnetic resonance imaging of mesenchymal stem cells in myocardial infarction. *Circulation* **107**, 2290–3.

53. Amsalem, Y., Mardor, Y., Feinberg, M. S., Landa, N., Miller, L., Daniels, D., et al. (2007) Iron-oxide labeling and outcome of transplanted mesenchymal stem cells in the infarcted myocardium. *Circulation* **116**, I138–45.

54. Barbash, I. M., Chouraqui, P., Baron, J., Feinberg, M. S., Etzion, S., Tessone, A., et al. (2003) Systemic delivery of bone marrow-derived mesenchymal stem cells to the infarcted myocardium: feasibility, cell migration, and body distribution. *Circulation* **108**, 863–8.

55. Bindslev, L., Haack-Sørensen, M., Bisgaard, K., Kragh, L., Mortensen, S., Hesse, B., et al. (2006) Labelling of human mesenchymal stem cells with indium-111 for SPECT imaging: effect on cell proliferation and differentiation. *Eur J Nucl Med Mol Imaging* **33**, 1171–7.

56. Brenner, W., Aicher, A., Eckey, T., Massoudi, S., Zuhayra, M., Koehl, U., et al. (2004) 111In-labeled CD34+ hematopoietic progenitor cells in a rat myocardial infarction model. *J Nucl Med* **45**, 512–8.

57. Liu, X.-H., Bai, C.-G., Xu, Z.-Y., Huang, S.-D., Yuan, Y., Gong, D.-J., et al. (2008) Therapeutic potential of angiogenin modified mesenchymal stem cells: Angiogenin improves mesenchymal stem cells survival under hypoxia and enhances vasculogenesis in myocardial infarction. *Microvasc Res* **76**, 23–30.

58. Vulliet, P. R., Greeley, M., Halloran, S. M., MacDonald, K. A., and Kittleson, M. D. (2004) Intra-coronary arterial injection of mesenchymal stromal cells and microinfarction in dogs. *Lancet* **363**, 783–4.

59. Breitbach, M., Bostani, T., Roell, W., Xia, Y., Dewald, O., Nygren, J. M., et al. (2007) Potential risks of bone marrow cell transplantation into infarcted hearts. *Blood* **110**, 1362–9.

60. Quevedo, H. C., Hatzistergos, K. E., Oskouei, B. N., Feigenbaum, G. S., Rodriguez, J. E., Valdes, D., Pattany, P. M., Zambrano, J. P., Hu, Q., McNiece, I., Heldman, A. W., Hare, J. M. (2009) Allogeneic mesenchymal stem cells restore cardiac function in chronic ischemic cardiomyopathy via trilineage differentiating capacity. *Proceedings of the National Academy of Sciences* **106**, 14022–7.

61. Schuleri, K. H., Feigenbaum, G. S., Centola, M., Weiss, E. S., Zimmet, J. M., Turney, J., Kellner, J., Zviman, M. M., Hatzistergos, K. E., Detrick, B., Conte, J. V., McNiece, I., Steenbergen, C., Lardo, A. C., Hare, J. M. (2009) Autologous mesenchymal stem cells produce reverse remodelling in chronic ischaemic cardiomyopathy. *European Heart Journal* **30**, 2722–32.

62. Hare, J. M., Traverse, J. H., Henry, T. D., Dib, N., Strumpf, R. K., Schulman, S. P., Gerstenblith, G., DeMaria, A. N., Denktas, A. E., Gammon, R. S., Hermiller, Jr., J. B., Reisman, M. A., Schaer, G. L., Sherman, W. A. (2009) Randomized Double-Blind, Placebo-Controlled, Dose-Escalation Study of Intravenous Adult Human Mesenchymal Stem Cells (Prochymal) After Acute Myocardial Infarction. *Journal of the American College of Cardiology* **54**, 2277–86.

Chapter 6

Methods for Human Embryonic Stem Cells Derived Cardiomyocytes Cultivation, Genetic Manipulation, and Transplantation

Gil Arbel, Oren Caspi, Irit Huber, Amira Gepstein, Michal Weiler-Sagie, and Lior Gepstein

Abstract

A decade has passed since the initial derivation of human embryonic stem cells (hESC). The ensuing years have witnessed a significant progress in the development of methodologies allowing cell cultivation, differentiation, genetic manipulation, and in vivo transplantation. Specifically, the potential to derive human cardiomyocytes from the hESC lines, which can be used for several basic and applied cardiovascular research areas including in the emerging field of cardiac regenerative medicine, attracted significant attention from the scientific community. This resulted in the development of protocols for the cultivation of hESC and their successful differentiation toward the cardiomyocyte lineage fate. In this chapter, we will describe in detail methods related to the cultivation, genetic manipulation, selection, and in vivo transplantation of hESC-derived cardiomyocytes.

Key words: Human embryonic stem cells, Genetic selection, Cell culture, Transplantation

1. Introduction

Human embryonic stem cells (hESC) are pluripotent cell lines that were isolated from the inner cell mass of human blastocysts (1). These unique cells are characterized by their capacity for prolonged undifferentiated proliferation in culture while maintaining the potential to differentiate into derivatives of all germ layers. In this chapter, we will focus on the differentiation of hESC to cardiomyocytes based on work conducted by our laboratory (2) and others (3–5). The cardiomyocyte differentiation system established in our laboratory (2) is based on the embryoid body (EB) differentiating system, where the hESC are initially cultivated as

Randall J. Lee (ed.), *Stem Cells for Myocardial Regeneration: Methods and Protocols*, Methods in Molecular Biology, vol. 660, DOI 10.1007/978-1-60761-705-1_6, © Springer Science+Business Media, LLC 2010

undifferentiated cells on mouse embryonic fibroblasts (MEF) feeder layer. In order to allow differentiation, the hESC are next grown in suspension as three dimensional cell aggregates and are later plated on gelatin-coated plates where they start to display spontaneously contracting areas. Cells isolated from the beating areas, within the EBs, demonstrate molecular, structural, and functional properties of early-stage human cardiomyocytes (2). Further studies from our group revealed a temporal pattern of hESC-derived cardiomyocyte (hESC–CMs) proliferation, cell-cycle withdrawal, and ultrastructural maturation (6). Successful application of these unique cells in the areas of cardiovascular research and regenerative medicine has been hampered by difficulties in identifying and selecting specific cardiac progenitor cells from the mixed population of differentiating cells. To overcome this obstacle, we recently generated stable transgenic lines using a lentiviral vector containing a cardiac specific promoter driving the expression of eGFP (7), enabling selection of cardiomyocytes by Fluorescence-Activated Cell Sorting (FACS). Recent studies have assessed the feasibility of hESC–CMs transplantation in the uninjured and infarcted rodent heart (8, 9). While different protocols were used in these studies to derive the hESC–CMs and transplant them in the infarcted heart, all of these studies demonstrated the ability of the hESC–CMs to survive within the infarcted area and favorably affect cardiac function.

2. Materials

2.1. Mouse Embryonic Feeder Layer Production

1. 5% Trypsin-EDTA (Invitrogen, Carlsbad, CA), is aliquoted and stored at –20°C.

2. MEF-P/S medium: 90% DMEM (Invitrogen), 10% Fetal Bovine Serum (Biological Industries, Bet-haemek, Israel) and 100 U/ml penicillin–100 µg/ml streptomycin(Biological Industries) (see Note 1).

3. MEF medium: 90% DMEM (Invitrogen), 10% Fetal Bovine Serum (Biological Industries) (see Note 1).

4. MEF Cryopreservative: 60% DMEM (Invitrogen), 20% defined Fetal Bovine Serum (HyClone, Logan, Utah, USA), and 20% DMSO (Sigma, St. Louis, Missouri, USA), is prepared on demand and filter sterilized using a 0.22 µm filter.

2.2. MEFs Cultivation

1. Mitomycin solution: 8 µg/ml mitomycin C (Sigma) in DMEM (Invitrogen). The solution is filter sterilized, aliquoted and stored at –20°C.

2. Gelatin-coated plates: Gelatin from porcine skin, type A (Sigma) is dissolved in sterile water (Sigma) at a concentration of 0.1% and autoclaved. A 2-ml gelatin solution per well is

added to a 6-wells plate, and incubated overnight at 37°C. The coated plates can be kept in the incubator for up to a month. The gelatin solution is aspirated before use.

3. ES serum-free medium: 79% Knock-Out DMEM, 20% Knock-Out Serum Replacement (SR), 1% nonessential amino acids, 1 mM l-glutamine, 0.1 mM β-mercaptoethanol, and 4 ng/ml bFGF (R&D Systems, Minneapolis, MN) (all from Invitrogen, except when indicated) (see Note 1).

2.3. Human Embryonic Stem Cells Cultivation

1. ES full medium: 79% Knock-Out DMEM, 20% defined Fetal Bovine Serum (HyClone), 1% nonessential amino acids, 1 mM l-glutamine, and 0.1 mM β-mercaptoethanol (all from Invitrogen, except when indicated) (see Note 1).

2. Fibronectin covered plates: Fibronectin 1 mg/ml (Biological Industries) is diluted 1:50 in PBS and filter sterilized. 1 ml fibronectin solution per well is added to a 6-wells plate and incubated at 37°C for 1 h. Coated plates can be stored at 4°C for 1 week. The fibronectin is aspirated before use.

3. Collagenase type IV (Wortington, Lakwood, NJ, USA) is dissolved at day of use in Knock-Out DMEM (Invitrogen) to a final concentration of 300 units/ml, and filter sterilized.

4. ES Cryopreservative: 60% Knock-Out DMEM (Invitrogen), 20% defined Fetal Bovine Serum (HyClone), and 20% DMSO (Sigma).

2.4. Embryoid Bodies Creation

1. EBs 20% medium: 79% DMEM, 20% defined Fetal Bovine Serum, 1% nonessential amino acids, 0.1 mM β-mercaptoethanol, and 100 U/ml penicillin–100 μg/ml streptomycin (Biological Industries) (all from Invitrogen, except when indicated) (see Note 1).

2. EBs 15% medium: 74% DMEM, 15% defined Fetal Bovine Serum, 1% nonessential amino acids, 0.1 mM β-mercaptoethanol, and 100 U/ml penicillin–100 μg/ml streptomycin (Biological Industries) (all from Invitrogen, except when indicated) (see Note 1).

2.5. Embryoid Bodies Dissociation

1. Low Ca solution: 120 mM NaCl, 5.4 mM KCl, 5 mM $MgSO_4$, 5 mM Na-pyruvate, 20 mM glucose, 20 mM taurine, and 10 mM HEPES (all from Sigma) at pH 6.9 (see Note 6).

2. CaCl solution: 3 mM CaCl (Sigma) (see Note 6).

3. Collagenase B solution: 3 mg collagenase B (Roche, Mannheim, Germany), 3 ml low Ca solution and 30 μl CaCl solution is prepared at day of use and filter sterilized.

4. Resuspension solution: 85 mM KCl, 30 mM K_2HPO_4, 5 mM $MgSO_4$, 1 mM EGTA, 2 mM Na_2-ATP, 5 mM Na-pyruvate, 5 mM creatine, 20 mM taurine, and 20 mM glucose (all from Sigma) at pH 7.2 (see Note 6).

2.6. Generation of Transgenic hESC Lines

1. Poly-l-Lysine (PLL) coated plates: 10-cm TC plates are coated with 6 ml of 0.1 mg/ml PLL (Sigma) for 1 h at room temperature, followed by solution aspiration and exposure to UV light. The plates are maintained at room temperature until used.

2. HEK medium: 90% DMEM (Invitrogen), 10% Fetal Bovine Serum (Biological Industries), and 100 U/ml penicillin–100μg/ml (Biological Industries) (see Note 1).

3. Human myosin light chain-2 V (MLC-2 V) promoter fragment: 560 bp of the MLC-2 V untranslated region, −513 to +47 related to the transcription initiation point. (Primers for PCR amplification of this fragment: Sense GCCACAGTGCCAGCCTTCATGG, antisense GTGGAAAGGACCCAGCACTGCC).

4. Lentiviral plasmids: 15 μg of the lentivirus vector containing the eGFP gene under the transcriptional control of the above mentioned MLC-2 V promoter fragment, 10 μg of the lentiviral packaging cassette expression plasmid (ΔNRF), and 5 μg of the VSV-G envelope expression plasmid.

5. Polybrene stock solution: polybrene (Sigma) is dissolved in PBS at a concentration of 1 mg/ml. The stock solution is stable at 4°C for 1 month.

6. 0.25% trypsin-EDTA solution (Biological Industries).

2.7. Transplantation of hESC-Derived Cardiomyocytes to the Uninjured and Infracted Rat Heart

1. Immunesuppression: Cyclosporine A (Novartis, Camberley, UK) 15 mg/kg and methylprednisolone (Pfizer, NY, NY) 2 mg/kg.

2. Anesthesia: Ketamine (Vetoquinol, Lure, france) 90 mg/kg and Xylazine (Kepro, Deventer, Holland) 10 mg/kg.

3. Analgesia: Ketoprofen (IVX Animal Health, Inc., St. Joseph, MO) 3–5 mg/kg/day.

4. Rodent ventilation: rodent ventilator (Harvard apparatus model 683,MA).

5. Iris Scissors (Delicate) sharp/blunt, Alm retractor, Castroviejo Needle holder, Graefe forceps (0.8 mm tip) (all from fine science tools (FST), foster, CA).

3. Methods

3.1. MEF Layer Production

1. MEFs are used as a feeder layer for hES cells. This feeder layer prevents ES differentiation and keeps them in a pluripotent state. The MEFs are obtained from pregnant imprinting control region (ICR) mice at day 13 of pregnancy.

2. The mice are sacrificed using CO_2. Uterus is removed, placed in a bacteriology Petri dish and washed with sterile PBS.

3. The embryos are released from the uterus using forceps and scissors. Placenta and membranes are removed, and the embryos are washed three times with 10 ml of PBS.

4. The tissue is minced using curved Iris scissors. Two milliliters of trypsin-EDTA is added and mincing is continued.

5. Five milliliters trypsin-EDTA is added and the tissue is incubated for 10 min at 37°C.

6. Trypsin is neutralized by 10 ml MEF-P/S medium and dish content is transferred to a 15 ml tube.

7. Debris is allowed to settle (without centrifugation) and the supernatant and debris are divided between several T75 culture flasks (3–5 embryos per flask). Twenty milliliters of MEF medium are added to each flask.

8. The flasks are cultured for 3 days, in which MEF medium is replaced at least once.

9. After 3 days, frozen stocks are prepared: Flasks are trypsinized for 10 min, neutralized with MEF medium and centrifuged at $200 \times g$ for 5 min. The pellet from each flask is resuspended in 1.5 ml of MEF medium. 1.5 ml of cryopreservative is added drop by drop. The cells from each flask are divided to three cryogenic vials that are frozen at –70°C in a freezing container (Nalgene, Rochester, NY) over night, and then transferred to liquid nitrogen.

3.2. MEFs Cultivation

1. MEFs are thawed rapidly by immersion in a 37°C water bath. The vial content is pipetted once up and down using a 1 ml pipette, and placed in a 15 ml tube. Two milliliters of MEF medium are added drop by drop. The tube is centrifuged at $200 \times g$ for 5 min. The pellet is resuspended in 10 ml of MEF medium and placed in T25 culture flask.

2. When cells reach confluence (usually after 5 days), they are trypsinized and diluted at ratio of 1:3–1:6 into larger flasks (75 cm²). MEFs can be diluted twice (three passages) before final plating.

3. Inactivation of MEFs – MEFs flask is washed with PBS and 6 ml of mitomycin solution is added to the cells. The flask is incubated at 37°C for 3 h. The mitomycin is aspirated and cells are trypsinized and counted.

4. Feeder layer plates for hES colonies are prepared by seeding of 300,000 cells per well on gelatin coated 6-wells plates, in MEF medium. The MEF covered plates can be kept in the incubator for 2 weeks.

5. MEF condition medium is prepared by seeding of 1.5×10^6 cells in 10 cm TC dish. The cells are cultivated at 10 ml of ES serum free medium. The condition medium is collected every 24 h for 5 days. It is filtered and can be stored at –20°C.

3.3. Human Embryonic Stem Cells Cultivation

1. hESC require MEF's secreted factors in order to remain pluripotent. These factors can be provided by growing the hES on plates coated with inactivated MEFs, or by culturing them with MEF condition medium. hESC that are grown on MEFs are cultured in ES medium, with serum or with serum replacement. hESC that are grown with MEF condition medium are seeded on fibronectin-covered plates . All plates are cultured at 2 ml medium per well, and the medium is replaced daily.

2. Cells are diluted when colonies become crowded, usually every 4–5 days. Each well is incubated with 0.5 ml of collagenase solution at 37°C for 45 min. At this time point, hES colonies usually detach from the plate while MEF cells still adhere (see Note 2).

3. One milliliter of ES medium is added to each well and colonies are vigorously pipetted in order to separate the clumps of cells.

4. Cells and medium are collected into test tubes and centrifuged at $150 \times g$ for 2 min. Supernatant is aspirated and the pellet is washed with 2 ml of ES medium and centrifuged again.

5. Cells are resuspended in ES medium and plated on MEF covered plates. Cells are seeded at a ratio of 1:3–1:4. Dilution should be determined by the growth rate of the specific ES line.

6. Cells can be frozen at the end of step 4 by resuspending the cells in ES full medium and slowly adding to it an equal volume of ES Cryopreservative. Cells from each well are divided into two cryogenic vials, frozen at –70°C in a freezing container (Nalgene, Rochester, NY) for 24 h, and then transferred to liquid nitrogen (see Note 3).

3.4. Embryoid Bodies Creation

1. In order to allow differentiation, hES cells must be allowed to form three-dimensional bodies, called embryoid bodies (EBs). The cells are treated with collagenase solution as described in Subheading 3.2, steps 1–4.

2. Cells from three wells should be collected giving rise to one suspension. The cells are resuspended in 9 ml ES full medium (see Note 3), and plated in 60 mm bacteriological petri dish for 8–10 days. After 4–5 days in suspension 5 ml of the medium is gently collected into a test tube and centrifuged at

$100 \times g$ for 1.5 min. The supernatant is aspirated and the pellet is resuspended in 5 ml of fresh medium and returned to the plate (see Note 4).

3. The EBs are plated after 8–10 days in suspension. The suspension is collected gently into a test tube and centrifuged at $100 \times g$ for 4.5 min. The EBs from each suspension are resuspended in 12 ml of EBs 20% medium and distributed evenly between the six wells of a gelatin-coated plate (see Note 5).

4. EBs 20% medium is replaced daily at days 2–7 from plating. From day 8 onward, 15% EBs medium is used and is replaced every other day.

5. Contracting areas within the EBs begin emerging from day 4 of plate seeding, and reach their maximal number at day 14–21.

3.5. Embryoid Bodies Dissociation

1. Contracting areas within the EBs are detected under the microscope, using 40× magnification. The location of these areas is marked on the bottom of the plate using a marker.

2. Microdissection of contracting areas is done under a binocular placed in a sterile hood. Two 23G needles are used for cutting out the EBs from the surrounding tissue. The extracted EBs are collected into a 60-mm bacteriological Petri dish containing EBs 20% medium and can be kept in an incubator for up to 48 h before dissociation.

3. The extracted EBs and medium are collected into a 15-ml sterile test tube and centrifuged at $400 \times g$ for 5 min.

4. The pellet is washed three times in 10 ml of PBS and centrifuged at $150 \times g$ for 2 min each time.

5. Collagenase B solution prewarmed to 37°C is added to the pellet. The tube is rotated at 37°C for 30–50 min (enzyme and cell batch dependent). The exact duration of incubation is determined by the size of the EBs. The incubation is terminated when small clumps of cells remain in the tube, but before complete dissociation. If cell clumps remain large after 50 min they can be pipetted a few times using a 1,000 μl tip and then incubated in rotation for another 10 min.

6. Cells are centrifuged at $400 \times g$ for 5 min The small cell clumps obtained can be used for animal transplantation without the use of resuspension solution.

7. For complete dissociation of the clumps to single cells, the pellet is resuspended in 3 ml of the resuspension solution and cells are incubated in rotation at 37°C for 15 min.

8. Cells are centrifuged at $400 \times g$ for 5 min and washed twice with 5 ml of EBs 20% medium.

**3.6. Generation
of Transgenic hESC
Line Containing
a Selectable Marker
for Cardiomyocytes**

The transgenic hESC line can express a reporter gene (eGFP) under the transcriptional control of a cardiac-specific promoter: the human myosin light chain-2 V promoter. This line is generated using a lentiviral system (see Note 7). This transgene-line derived cardiomyocytes express eGFP and can be identified and sorted using FACS.

1. Human embryonic kidney (HEK) 293 T cells are seeded on PLL-coated plates at 50–80% confluence one day prior to transfection.

2. One hour before transfection, medium is replaced to 5 ml of fresh HEK Medium. The cells are transfected with the lentiviral plasmids using the conventional calcium phosphate transient transfection method.

3. Five hours after transfection, medium is replaced with 10 ml of fresh HEK Medium. Twenty six hours after transfection, medium is replaced with 5 ml of fresh HEK Medium. Fifty five hours after transfection, the viral particles containing medium are collected. Five milliliters of fresh medium is added to the HEK cells (see Note 8).

4. The collected medium is centrifuged at $600 \times g$ for 7 min.

5. Supernatant is filtered through a 45-μm filter and then concentrated using Vivaspin, membrane cut-off 100,000 (Vivascience, Goettingen, Germany).

6. The concentrated virus particles containing media are supplemented with 6 μg/ml polybrene and added to ES full medium (see Note 9).

7. hECS are infected as clumps of ~200 cells obtained at the time of routine passage and are seeded on top of MEF-covered plate in the infection medium.

8. Sixteen hours after infection, the viral particles are collected again and the same hESC are infected for the second time (as described at stages 7–10) without passaging of the cells.

9. Seven to twelve hours after the second infection, medium is replaced with fresh ES full medium.

10. Seventy two hours after the second infection, eGFP expression can be detected.

11. Colonies that demonstrate robust eGFP expression are dissected under the microscope, transferred to separate wells and are continuously cultured. This step is repeated two to three times to generate a stable, homogenous line.

12. To generate single cell clones, the transgenic hESC colonies established can be digested using trypsin for 10 min. Cells are then counted and plated in MEF-covered 24-wells plates at a concentration of a single cell/well. The single cell-derived clones are continuously cultured, and the ones demonstrating

robust, stable, long-term, and homogenous expression of the transgene are chosen for propagation.

13. FACS sorting-EBs are detached from the plate using trypsin for 1 min, and then dissociated into single cells using collagenase B as described in the EBs dissociation protocol. Cells are then filtered through 40 μm nylon cell strainer (BD Biosciences, Bedford, MA) and resuspended in EBs 20% medium in concentration of 10^6 cells/ml. Flow cytometric analysis is preformed using a FACS sorter. A530/30 nm bandpass filter is used to measure eGFP fluorescence intensity excited with the 488 nm line of an argon ion laser. Detector settings are calibrated with untransfected hESC-derived EBs that are digested by the same method. The FACS-sorted cells are plated on gelatin coated 24-wells culture plates at a density of 10^5 cells/well.

3.7. Transplantation of hESC-Derived Cardiomyocytes to the Uninjured and Infracted Rat Heart

1. Sprague–Dawley rats are treated 24 h prior to cell transplantation by subcutaneous injection of Cyclosporine-A and methylprednisolone to allow immunosupression (see Note 10).

2. Prior to operation, the rats are anesthezied using an intramuscular injection of Ketamine and Xylazine and analgesia Ketoprofen (5 mg/kg) should be given subcutaneously (see Note 11).

3. Following anesthesia induction, the rat's chest and left axillary region are gently shaved.

4. To allow intubation, rats are positioned on top of a warm platform (37°C) in a position allowing neck extension.

5. Intubation is performed using a 22G intravenous cannula (with a blunt end) and a mini laryngoscope to allow elevation of the tongue base and hence proper visualization. The vocal cords should be located using lighting directed on the rat's neck.

6. Mechanical ventilation using a rodent ventilator (Harvard apparatus model 683, MA) is set at a rate of 75–85 breaths per minute with a tidal volume of 7 ml/kg.

7. Following chest sterilization using ethanol (70%) and iodine solution, skin incision should be conducted from the left axilla to the area of the xiphoid using blunt/sharp scissors and blunt forceps.

8. The skin and subcutaneous tissue are separated from the underneath muscles. Next, the pectoralis major and minor muscles should be separated and retracted to allow chest wall exposure.

9. Following Chest wall exposure and muscle retraction the chest wall should be opened carefully using blunt forceps avoiding damage to surrounding viscera.

10. Ribs are retracted with alm retractor.

11. Pericard is removed with blunt forceps or applicator.

12. *To induce myocardial infarction,* ligate the left anterior descending artery located at the cross of the vertical imaginary line connecting the apex with the left atrium – pulmonary artery border and the horizontal imaginary line just below the inferior border of the left atrium. Ligation should be conducted using 6–0 Prolene suture handled by a Castroviejo needle holder to create a double stitch and then two separate single stitches.

13. *To inject cells* – Inject cells 0.5–3×10^6 in serum free medium 100–200 mL. Cells should be injected using a curved needle (29–30G) following cell dissociation. Proper injection is characterized by the appearance of a white halo in the area of injection.

14. Following the target procedure (Myocardial infarction/Cell injection), the chest wall should be closed carefully with Vircyl 3–0 sutures. Skin should be sutured using a Nylon 3–0 nylon sutures.

4. Notes

1. All media are filter sterilized and stable at 4°C for up to 1 month.

2. Alternatively, colonies can be incubated with the collagenase for 10 min and then scraped using a cell scraper or a pipette. Using this method results in the removal of the MEF cells as well as the hESC.

3. Both cells that were grown with serum and with SR require ES full medium, containing serum, for this step.

4. The EBs should be floating in the suspension medium. If they adhere to the bottom of the plate, try to use a different brand of petri dishes.

5. EBs are susceptible to mechanical stress and therefore should be mixed, pipetted and centrifuged gently.

6. All solutions are filter sterilized and kept at 4°C.

7. Safety measurements for handling lentiviral infection:

    (a) All work is done in a level-2 biohazard hood, using protective equipment including a long sleeved lab coat, two sets of gloves, goggles, and respiratory protection.

    (b) Waste medium, tips, and filters are collected into a container containing 1:10 diluted bleach. Waste pipettes are washed with diluted bleach.

(c) All waste is disposed in a double biohazard bag.

(d) Surfaces and tools are cleaned with 1:10 diluted bleach and exposed to U.V. light at the end of work.

8. Transfection efficiency is examined 48 h following cell transfection. If less than 30% of the HEK cells express GFP, the transfection condition should be altered.

9. The virus particles containing media should be concentrated 5–20 times for ES infection. The precise concentration is chosen based on the transfection efficiency.

10. Following cell injection, rats are treated with subcutaneous injection of Cyclosporine-A (15 mg/kg/day) and methylprednisolone (2 mg/kg/day) to prevent graft rejection and allow proper immunosuppression.

11. Analgesia should be given in the initial 3 days following surgery at a dosage of 3 mg/kg (Ketoprofen).

References

1. Thomson, J. A., Itskovitz-Eldor, J., Shapiro, S. S., Waknitz, M. A., Swiergiel, J. J., Marshall, V. S., and Jones, J. M. (1998) Embryonic stem cell lines derived from human blastocysts, *Science 282*, 1145–1147.

2. Kehat, I., Kenyagin-Karsenti, D., Snir, M., Segev, H., Amit, M., Gepstein, A., Livne, E., Binah, O., Itskovitz-Eldor, J., and Gepstein, L. (2001) Human embryonic stem cells can differentiate into myocytes with structural and functional properties of cardiomyocytes, *J Clin Invest 108*, 407–414.

3. Xu, C., Inokuma, M. S., Denham, J., Golds, K., Kundu, P., Gold, J. D., and Carpenter, M. K. (2001) Feeder-free growth of undifferentiated human embryonic stem cells, *Nat Biotechnol 19*, 971–974.

4. Mummery, C., Ward-van Oostwaard, D., Doevendans, P., Spijker, R., van den Brink, S., Hassink, R., van der Heyden, M., Opthof, T., Pera, M., de la Riviere, A. B., Passier, R., and Tertoolen, L. (2003) Differentiation of human embryonic stem cells to cardiomyocytes: role of coculture with visceral endoderm-like cells, *Circulation 107*, 2733–2740.

5. Cao, F., Wagner, R. A., Wilson, K. D., Xie, X., Fu, J. D., Drukker, M., Lee, A., Li, R. A., Gambhir, S. S., Weissman, I. L., Robbins, R. C., and Wu, J. C. (2008) Transcriptional and functional profiling of human embryonic stem cell-derived cardiomyocytes, *PLoS One 3*, e3474.

6. Snir, M., Kehat, I., Gepstein, A., Coleman, R., Itskovitz-Eldor, J., Livne, E., and Gepstein, L. (2003) Assessment of the ultrastructural and proliferative properties of human embryonic stem cell-derived cardiomyocytes, *Am J Physiol Heart Circ Physiol 285*, H2355–H2363.

7. Huber, I., Itzhaki, I., Caspi, O., Arbel, G., Tzukerman, M., Gepstein, A., Habib, M., Yankelson, L., Kehat, I., and Gepstein, L. (2007) Identification and selection of cardiomyocytes during human embryonic stem cell differentiation, *FASEB J 21*, 2551–2563.

8. Caspi, O., Huber, I., Kehat, I., Habib, M., Arbel, G., Gepstein, A., Yankelson, L., Aronson, D., Beyar, R., and Gepstein, L. (2007) Transplantation of human embryonic stem cell-derived cardiomyocytes improves myocardial performance in infarcted rat hearts, *J Am Coll Cardiol 50*, 1884–1893.

9. Laflamme, M. A., Chen, K. Y., Naumova, A. V., Muskheli, V., Fugate, J. A., Dupras, S. K., Reinecke, H., Xu, C., Hassanipour, M., Police, S., O'Sullivan, C., Collins, L., Chen, Y., Minami, E., Gill, E. A., Ueno, S., Yuan, C., Gold, J., and Murry, C. E. (2007) Cardiomyocytes derived from human embryonic stem cells in pro-survival factors enhance function of infarcted rat hearts, *Nat Biotechnol 25*, 1015–1024.

Part II

Animal Models for Assessing Stem Cells for Myocardial Regeneration

Chapter 7

Experimental Cell Transplantation Therapy in Rat Myocardial Infarction Model Including Nude Rat Preparation

Wangde Dai and Robert A. Kloner

Abstract

As a novel potential therapeutic strategy for cardiac disease, cell transplantation therapy has been extensively investigated in experimental studies and clinical trials. Although encouraging results have been demonstrated, a number of critical questions still remain to be answered. For example, what kind of stem cell and how many cells should be used; what is the best time for cell transplantation after acute myocardial infarction; which delivery approach is better, intravenous injection or direct intramyocardial injection? Transplantation of cells derived from human tissues into experimental animals may elicit an immune rejection. Immunodeficient nude rats provide a useful myocardial infarction model for cell transplantation therapy studies. We introduce our detailed methods of direct intramyocardial injection of immature heart cells and stem cells into the myocardial infarction region of rats and nude rats. Careful maintenance under aseptic conditions and proper surgical technique are essential to improve the survival of immunodeficient rats after surgery.

Key words: Cell transplantation therapy, Myocardial infarction, Stem cell, Nude rat

1. Introduction

Experimental and clinical studies have demonstrated that stem cell transplantation therapy provides a novel alternative approach for the repair of damaged myocardium. However, many questions remain to be resolved before cell therapy becomes common general clinical practice for the diseased heart (for review, see ref. 1–3). For example, the search for the best donor cell source is still an ongoing process. Different kinds of stem cells, such as embryonic stem cells, resident cardiac stem cells, bone marrow-derived cells, human umbilical cord blood stem cells, adipose tissue-derived stem

Randall J. Lee (ed.), *Stem Cells for Myocardial Regeneration: Methods and Protocols*, Methods in Molecular Biology, vol. 660, DOI 10.1007/978-1-60761-705-1_7, © Springer Science+Business Media, LLC 2010

cells, and skeletal muscle-derived stem cells, have been used for cardiac regenerative studies. Although encouraging results have been demonstrated, published data indicated that transplanted stem cells may form teratomas, induce arrhythmias, elicit an immune rejection, fail to differentiate into cardiac cells, or may not integrate with the recipient myocardium and lack electromechanical coupling with the neighboring host myocardium. Washout of transplanted cells through the vascular system of the heart (4), or low survival rate of transplanted cells due to the poor supply of oxygen and nutrients within the ischemic area also limit the efficiency of cell therapy for heart disease (1–3). Therefore, additional experimental studies are needed to overcome these hurdles for stem cell transplantation therapy in heart.

Immunodeficient nude rats subjected to myocardial ischemia/reperfusion provide an important research model for cell transplantation studies. Our research group (5) transplanted human embryonic stem cell (hESC)-derived cardiomyocytes into hearts subjected to ischemia/reperfusion in immunodeficient nude rats and immune competent Sprague–Dawley rats. Four weeks later, the transplanted cells injected into nude rats survived, formed sizable grafts, expressed cardiac muscle markers, exhibited sarcomeric structure, and were well interspersed with the endogenous myocardium (Fig. 1). However, in immune competent Sprague–Dawley rats, these donor cells induced an overt lymphocytic infiltrate in the injection area, and did not escape immune surveillance. In this chapter, we describe how to make the rat myocardial infarction model and how to perform direct intramyocardial injection of cells, as well as some useful measurements to assess the efficiency of cell transplantation therapy in rats. We also describe the additional precautions that are needed when working with nude rats in this experimental model of myocardial ischemia.

2. Materials

2.1. Equipment

1. Immune competent rats (Sprague–Dawley rats or equivalent) and immunodeficient nude rats (Charles River Laboratories, Inc., Wilmington, MA).

2. Operating table.

3. Harvard rodent ventilator (Model 683, Harvard Apparatus, South Natick, MA, USA) and intubation tube (PE200, Becton Dickinson and Company, Sparks, MD, USA).

4. MILLEX-GS 0.22 μm syringe driven filter unit (Millipore corporation, Bedford, MA, USA).

5. Small flat spatula for intubation.

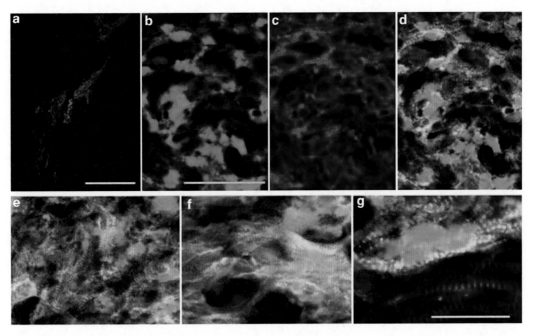

Fig. 1. Analysis of hESC-derived cardiomyocytes 4 weeks after transplantation into ischemic myocardium. (**a**) Low power image of GFP epi-fluorescence (*green*) in a heart engrafted with hESC-derived cardiomyocytes. (**b–d**) GFP epi-fluorescence (*green; panel B*), alpha-actinin immune reactivity (*red*, rhodamine-conjugated secondary antibody; *panel C*) and a merged image showing both signals (*panel D*). Sarcomeric organization in the cardiomyocytes shown is somewhat immature. (**e**) Merged image of GFP epi-fluorescence (*green*) and alpha-actinin immune reactivity (*red*) of cardiomyocytes with intermediate sarcomeric organization. (**f**) Merged image GFP epi-fluorescence (*green*) and alpha-actinin immune reactivity (*red*) of cardiomyocytes with a more mature sarcomeric organization. (**g**) Merged image GFP epi-fluorescence (*green*) and alpha-actinin immune reactivity (*red*) at the graft-host myocardium border. Host cardiomyocytes show alpha-actinin signal only (*red, bottom of the panel*). *Magnification bars* correspond to 1 mm (*panel A*), 50 μm (*panels B–F*) or 25 μm (*panel G*). (Originally published in ref. *5*. Reprinted with permission.)

6. Surgical instruments.

7. Sterile surgical gloves (Cardinal Health, McGaw park, IL, USA).

8. Sterile nonabsorbable suture (4-0; Ethicon, Somerville, NJ, USA).

9. Insulin syringe with permanently attached needle (28 gauge, Sherwood, St. Louis, MO, USA).

10. Sterile absorbable suture (2-0 PDSII; Ethicon).

11. Sterile skin staples (Michel clips 7.5 mm; Miller Instruments Company, Bethpage, NY, USA).

12. Heating pad (one for surgery and one for recovery).

13. Sterile laminar flow hood for the preparation of stem cells.

14. Sterile laminar flow hood for animal handling and surgery.

15. Cell counter.

16. Standard tabletop centrifuge.

17. Trypan blue dye (Gibco BRL, Grand Island, NY, USA).

18. Sterile Eppendorf tubes.

19. Vacuum steam sterilizer (Model 533LS, Getinge).

20. Millar MPVS-300 system (Millar instruments, Houston, TX, USA).

21. XISCAN 1000 X-ray system (XI TEC INC., Windsor Locks, CT, USA).

2.2. Reagents

1. Ketamine HCl (100 mg/ml, Abbott Laboratories, North Chicago, IL, USA).

2. Xylazine sterile solution (20 mg/ml, Ben Venue Laboratories, Bedford, OH, USA).

3. Chlorhexidine scrub/Chlorhexidine solution (First priority, INC., Elgin, IL, USA).

4. 70% Ethyl alcohol (Fisher Scientific, Rochester, NY, USA).

5. Bupivicaine (2.5 mg/ml, AstraZeneca LP, Wilmington, DE, USA).

6. Buprenorphine (0.3 mg/ml, Reckitt Benckiser Pharmaceuticals, Richmond, VA, USA).

7. Sterile 0.9% Sodium Chloride (Baxter, Deerfield, IL, USA).

3. Methods

3.1. Preparation of Myocardial Infarction Model in Immune Competent Rats

1. Weigh the rat. Anesthetize the rat with intraperitoneal injection of ketamine (75 mg/kg) and xylazine (5 mg/kg). Then shave the chest area.

2. Place the rat into the supine position. Insert the plastic tube into the trachea and mechanically ventilate with room air (rate 60 cycles/min, tidal volume 1 ml per 100 g body weight).

3. Make a skin incision from the base of the sternum toward the armpit. Free the skin from the underlying tissue using blunt dissection.

4. Dissect between the muscles of the chest to expose the ribs. Retract the upper layer of pectoralis major muscle back toward the sternum, and the next layer of pectoralis minor muscle toward the opposite direction using a suture. After exposing the ribs, inject bupivacaine (1 mg/kg body weight) into the third, fourth, and sixth intercostal muscles.

5. Use blunt dissection to make a hole in the fourth intercostal space into the chest, and insert a retractor into the chest cavity through this hole. Note that the teeth of the retractor

should face up upon entering the chest cavity. Then rotate the retractor toward the head of the rat to avoid trapping the lungs in the teeth of the retractor.

6. Widen the retractor to expose the heart. Remove the pericardium.

7. Locate the coronary artery (see Note 1) and place a 4–0 silk suture around the artery. Ligate the artery by tying the suture in a double surgeons' knot. The color of the anterior free wall of left ventricle changes from pink to pale after successful ligation of the coronary artery (see Note 2). Cut the suture and leave approximately 0.5 cm for holding the heart in position during cell injection.

8. Check that there is no bleeding in the chest cavity and remove the retractor. Close the opened fourth intercostal space by suturing with absorbable suture. Be careful not to injure the lung with the needle when suturing. In order to remove the air from the chest, gently squeeze the chest while tying the last suture. Replace the retracted chest muscle. Close the chest skin with sterile stainless staples. Inject subcutaneously normal saline (1 ml/100 g body weight). Place the rat on a heating pad and monitor until it fully recovers. Give Buprenex (0.001 mg/100 g body weight, subcutaneous injection twice daily) for 2 days as analgesic.

3.2. Cell Injection

1. At 1 week after myocardial infarction, prepare the rat following the same procedures as outlined in Subheading 3.1, steps 1–6. Expose the heart through the fifth intercostal space. The infarcted heart may adhere to the chest wall due to the first surgery. Free the adhered heart from the chest wall with the tip of a sterile Q-tip.

2. Expose the heart. The myocardial infarct scar should appear pale, grey in color, and should not be contracting. The scar should be easily distinguishable from the surrounding normal thick, pink contracting muscle. In some cases, the suture for the coronary occlusion from surgery can be held with a forceps to stabilizing the heart for cell injection.

3. Bend the distal end of the needle of a 28 gauge insulin syringe at a 45° angle for injecting the cells. Enter the left ventricular wall with the end of the needle parallel to the epicardial plane. The entering site should be a short distance from the edge of infarction. Slowly advance the needle into the center of the infarcted region. The needle tip can be visualized under the epicardial layer. Be careful not to let the tip penetrate into the LV cavity or perforate outside of the heart through the epicardium. The formation of a pale bleb under the epicardium can be clearly visible when the injection is successful.

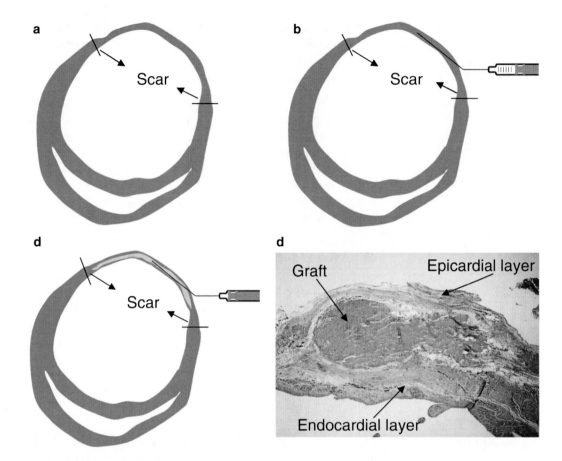

Fig. 2. (**a**) Schematic drawing of the heart with a thin scar in the anterior free wall after myocardial infarction in the rat. (**b**) We bend the top of the insulin needle and place it within the scar of the anterior free wall. (**c**) Schematic drawing shows the sandwiched engrafted cells (*light grey area* with the scar) between the endocardial and epicardial layer of the infarct scar at the end of injection. (**d**) Representative hematoxylin and eosin staining picture (magnification ×40) shows that transplanted rat neonatal cardiomyocytes form sizable graft (*boxed area by dotted yellow line*) between the endocardial and epicardial layer within the infarcted scar of the rat heart at 6 months after cell transplantation (*see* ref. *7*. Reprinted with permission.).

Inject about 70 µl volume into the infarcted zone of the heart in 150–200 g rat (see Note 3, Fig. 2).

4. After cell injection, close the chest and allow the animal to recover under the same guidelines as the first surgery.

3.3. Useful Parameters to Measure (Endpoints)

At the end of the experimental procedures, reanesthetize the rats using the same procedures listed above. In our laboratory, we often perform the following assessments:

1. Hemodynamics: Expose the right carotid artery and make a small incision on the artery wall. Insert a 2F high-fidelity, catheter-tipped micromanometer (model SPR-869, Millar, Inc) into the carotid artery. Advance the catheter into the ascending aorta to record heart rate and blood pressure.

Then, advance the catheter further into the LV to record LV hemodynamic parameters, including $\pm$ dP/dt, LV end systolic, and diastolic pressure.

2. Left ventriculography: After recording hemodynamic parameters, perform a left ventriculography to assess LV function with a XiScan 1000 C-arm X-ray system (XiTec, Inc; 3-in. field of view). Following intravenous injection of 1 ml non-ionic contrast through the left jugular vein, record video images of anterior–posterior and lateral projections on half-inch super-VHS videotape at 30 frames per second under constant fluoroscopy. Calculate systolic and diastolic LV volumes from the video images. All parameters are averaged over three consecutive cycles in both projections. LV Ejection fraction (%) is calculated as [$100 \times$ (volume in diastole – volume in systole)/volume in diastole], and averaged over both projections.

3. Assessment of LV regional wall motion: Assess the regional wall motion by left ventriculography. With the base of the heart as a reference, superimpose tracings of the LV circumference during end-diastole and end-systole of the same cardiac cycle on transparent film in both the anterior–posterior and lateral views. Measure LV bulging and akinesis. LV bulging occurs when the tracing from end-systole is not confined within that of end-diastole. When the tracing from end-systole is superimposable upon the tracing for end-diastole, then akinesis is present. Measure the length of total LV diastolic circumference, and circumferential length of the bulging or akinetic segment with computerized planimetry. The size of paradoxical LV systolic bulging (dyskinesis) or akinetic motion is expressed as percentage of total LV diastolic circumference.

4. Postmortem LV volumes: After assessing the in vivo cardiac function, euthanize the rats under deep anesthesia by injection of 1 ml potassium chloride (149 mg/ml) given through the left jugular vein. Excise the hearts and pressure-fix the hearts with formalin (pressure equal to 13 cm water column). Measure the postmortem LV volumes by filling the cavity with water and weighing (repeated three times and averaged).

5. Remodeling parameters: Cut the hearts into four transverse slices after LV volume measurement. Embed the slices in paraffin and process for histology. Stain the sections (5 µm thickness) with hematoxylin and eosin, and picrosirius red. Use computerized planimetry of the histological images of the stained sections to measure (1) scar thickness (average of five equidistant measurements); (2) septum thickness (average of three equidistant measurements); (3) the length of LV epicardial circumference and endocardial circumference;

(4) the length of circumference occupied by infarcted wall; (5) LV cavity area; (6) total LV area. The infarct size is expressed as percentage of total LV circumference. Expansion index, as defined by Hochman and Choo (6), is expressed as: [LV cavity area/total LV area×septum thickness/scar thickness].

6. Tracking the engrafted cells: The transplanted cells can be identified by different labeling. For example, if the transplanted cells express green fluorescent protein (GFP), recipient hearts are processed for frozen section. Engrafted cells show GFP epi-fluorescence under fluorescence microscopy.

7. Immunohistochemical staining: In order to characterize the grafted hESC-derived cardiomyocytes, immunohistochemical staining is performed on frozen sections of hearts. Various primary antibodies are used, including antibody against GFP (1:50, Chemicon), sarcomeric actin (1:75, Dako), myosin light chain-2a (MLC-2a; 1:50, Synaptic Systems), tropomyosin (1:1,000, Sigma), alpha myosin heavy chain (αMHC, 1:200, Hybridoma Bank, Iowa), alpha-actinin (1:800, Sigma), and connexin43 (1:200, Chemicon). Rhodamine-conjugated secondary antibodies are used to demonstrate the expression of tested markers as red color. The slides are viewed under confocal microscopy to show the colocalization of GFP epi-fluorescence (green) and Rhodamine fluorescence (red). Examples of signals are shown in Fig. 1.

3.4. Special Considerations When Using the Nude Rat in a Myocardial Infarction Model

Nude rats do not have a functional immune system. Therefore, the care of nude rats is very different from the care for immune competent rats (see Note 4). In order to reduce the risk of infection, everything that nude rats come into contact with must be autoclaved, including water, bedding, cage, etc. Static microisolator cages that contain an air filter are used to house the nude rats. During the autoclaving of the wrapped cages with bedding inside, the drying time needs be long enough to allow the inside bedding to become dry. In our laboratory, the cages are autoclaved twice a week. The food is irradiated (Harlan Teklad, Madison, WI, USA). Nude rats are housed in a separate room from other animals, and must be handled inside a laminar flow hood. The person handing the nude rats must wear a cap, mask, sterile gloves, and lab coat. In our laboratory, surgery on nude rats is performed in a sterile laminar flow hood as the following steps:

1. Preparation of laminar flow hood: Clear the working surface and completely sanitize the surface of the hood (including both sides, back, top, bottom, and the glass in front) with 70% alcohol. Wipe surfaces with sterilized paper towels.

2. Put on surgical cap, face mask, disposable surgical gown with cuffed sleeves and gloves. Make sure the gloves cover the cuff of the sleeve.

3. Spray the surface of the ventilator, a scale, heating pad, and surgical board with 70% alcohol and place in the laminar flow hood.

4. Spray the outside of the cage with 70% alcohol and place into the flow hood.

5. Remove filter top and wire bar lid from the cage and place beside the cage in the laminar flow hood.

6. In the sterile laminar hood, insert sterile plastic tube into the trachea and mechanically ventilate with filtered room air (see Note 5, Fig. 3). The myocardial infarct model is made following the same procedures as outlined in Subheading 3.2, steps 2–9.

7. For cell injection, open the chest and expose the heart following the same procedures as outlined in Subheading 3.4, steps 1–6. After the infarcted area is well exposed, inject cells into infarcted region following the same procedures as outlined in Subheading 3.2, steps 1–4.

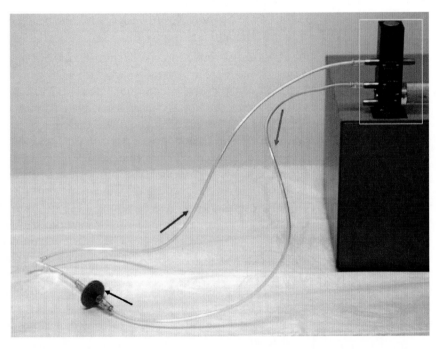

Fig. 3. This picture shows the plastic tubes connection between the ventilator and intubation tube. A 0.22-µm filter (*black arrow*) is put in the connecting tube from the ventilator to animal (*red arrow*). Blue arrow points to the connecting tube for exhaust air from animal. Because the *yellow boxed area* of the ventilator cannot be sterilized, the filter is helpful to prevent lung infection.

4. Notes

1. How to find the coronary artery? The left coronary artery is located at the interventricular groove (between the left and right ventricles). Its parallel veins appear dark red and are visible on the surface of the heart. The coronary artery appears as a thin white line under light, but in some rats it is not visible because it is intramural. The dark red colored veins are useful guides to determine the location of the coronary artery. Since the course of the veins parallels that of the arteries. The suture should be placed just below the left atrial appendage. Ligation of the left coronary artery at a more proximal (higher) position will result in a huge myocardial infarction and will increase postoperative mortality. Successful ligation of the coronary artery will result in color change (pale to cyanotic) on the surface of the free left ventricular wall.

2. What to expect after ligation of coronary? Successful ligation results in necrosis of the anterior free wall, evidenced by a pale, thin, and fibrotic scar distal to the suture. Within 24–48 h after a permanent coronary artery ligation, necrosis is fully developed. After successful ligation, some rats die because of either ventricular arrhythmias or congestive heart failure. If the ligation was not successful, no necrosis will be evident in the anterior free wall.

3. Intramyocardial injection: We choose an insulin syringe because it has very little dead space with a 28 gauge needle. Bending the top of the injection needle at a 45° angle is helpful to visualize the needle tip as it is advanced into the ventricular wall. Advance the needle from the edge of the infarct toward the central portion of the infarct, and draw back before injection to make sure that the needle does not penetrate into the ventricular cavity or blood vessels. If the needle has passed into the ventricle, remove it and start anew. Bleeding resulting from needle removal can be controlled by gentle pressure with a sterile Q-tip. Formation of a bleb (or a blister) on the surface of the heart is visible when injection is successful. Inject slowly and do not let excess fluid enter normal, nonischemic tissue as we have observed an increase in operative death when this occurs. The needle should be slowly removed in a gentle fashion to avoid tearing tissue and also to minimize loss of cells because of retrograde leaking from the injection site.

4. Every one touching the nude rats should be aware that nude rats do not have a functional immune system. Maintenance of nude rats in an aseptic environment is the key factor to reduce the mortality of nude rats.

5. Preparation of intubation: The intubation tube and connective plastic tube are immersed in 70% alcohol. Before intubation, the tubes are washed with sterile distilled water. There is a 0.22 μm syringe driven filter unit put in the connecting tube to prevent lung infection (Fig. 3).

Acknowledgment

This work was supported in part by grant from the National Institutes of Health (R01-HL073709), the Los Angeles Thoracic and Cardiovascular Foundation, and ES Cell International Pte Ltd in Singapore.

References

1. Dai W, Hale SL, Kloner RA. (2005) Stem cell transplantation for the treatment of myocardial infarction. *Transpl Immunol.* 15(2):91–7.

2. Dai W, Kloner RA. (2006) Myocardial regeneration by embryonic stem cell transplantation: present and future trends. *Expert Rev Cardiovasc Ther.* 4(3):375–83.

3. Dai W, Kloner RA. Mesenchymal stem cell therapy for the injured heart. In: Wollert KC, Field LJ, editors. Rebuilding the infarcted heart. London, UK: Informa Healthcare; 2007, p. 55–72.

4. Dow J, Simkhovich BZ, Kedes L, Kloner RA. (2005) Washout of transplanted cells from the heart: a potential new hurdle for cell transplantation therapy. *Cardiovasc Res.* 67(2):301–7.

5. Dai W, Field LJ, Rubart M, Reuter S, Hale SL, Zweigerdt R, Graichen RE, Kay GL, Jyrala AJ, Colman A, Davidson BP, Pera M, Kloner RA. (2007) Survival and maturation of human embryonic stem cell-derived cardiomyocytes in rat hearts. *J Mol Cell Cardiol.* 43(4):504–16.

6. Hochman JS, Choo H. (1987) Limitation of myocardial infarct expansion by reperfusion independent of myocardial salvage. *Circulation.* 75(1):299–306.

7. Muller-Ehmsen J, Peterson KL, Kedes L, Whittaker P, Dow JS, Long TI, Laird PW, Kloner RA. (2002) Rebuilding a damaged heart: long-term survival of transplanted neonatal rat cardiomyocytes after myocardial infarction and effect on cardiac function. *Circulation.* 105(14):1720–6.

Chapter 8

Large Animal Model of Heart Failure for Assessment of Stem Cells

Sharad Rastogi

Abstract

The field of stem cell biology and regenerative medicine is rapidly moving toward translation to clinical practice, and in doing so has become more dependent on animal donors and hosts for generating cellular reagents and assaying their potential therapeutic efficacy in models of human disease. Animal models of cardiovascular disease have proved critically important for the discovery of pathophysiological mechanisms and for the advancement of diagnosis and therapy. They offer a number of advantages; principally the availability of adequate healthy controls and the absence of confounding factors such as marked differences in age, concomitant pathologies, and pharmacological treatments. Over the past 30 years, investigators have developed numerous small and large animal models to study heart failure (HF). However, to translate discoveries from basic science into medical applications, research in large animal models becomes a necessary step. Intracoronary microembolizations-induced HF in dogs is an excellent large animal model of congestive HF for the assessment of pharmacological drugs, medical devices, and stem cells.

Key words: Stem cells, Intracoronary microembolizations, Heart failure

1. Introduction

Congestive heart failure (CHF) is a clinical syndrome in which pathophysiological underpinnings include left ventricular (LV) dysfunction, remodeling, and increased neurohumoral activation. CHF is a major focus of medical research. Its incidence has greatly increased in recent decades because of an aging population base and increasingly successful treatment of other forms of chronic cardiac disease.

When basic science discoveries are ready to be translated into medical applications, small animal models can be utilized as extremely useful tools for high throughput screenings. However, rodents are genetically very distant from humans and some

Randall J. Lee (ed.), *Stem Cells for Myocardial Regeneration: Methods and Protocols*, Methods in Molecular Biology, vol. 660, DOI 10.1007/978-1-60761-705-1_8, © Springer Science+Business Media, LLC 2010

pathophysiological features of certain disease and their response to pharmacological treatments may not be reliable predictors. Moreover, innovative interventions such as gene and cell therapy may pose completely different challenges when tested in large animals rather than small animals. For these reasons, dogs, pigs, monkeys, and sheep have been extensively used in cardiovascular research to generate preclinical models of HF. In general, dog and other large animal models of HF allow the study of LV function and volumes more accurately than the rodent models (1). In particular, they better allow chronic instrumentation. Furthermore, in dog myocardium like humans, the β-myosin heavy chain isoform predominates and excitation–contraction coupling processes are similar to the human myocardium (1, 2). There are several models of HF in dogs that include chronic rapid pacing, volume overload, transmyocardial direct-current shock, coronary artery ligation, and microembolization models. Intracoronary microembolizations-induced HF in dogs is an excellent large animal model of CHF for the assessment of pharmacological drugs, medical devices, and stem cell assessment, differentiation, mobilization, and/or functional improvement. The greatest advantage of this model is the ability to generate varying degrees of LV dysfunction by the number of embolic procedures performed.

In this chapter, we are going to describe how HF is produced in dogs by intracoronary microembolizations, and the material and methods used for isolation of bone marrow stem cells (BMSC) from blood, measurement of circulating Sca1-positive BMSC, measurement of mRNA, and protein expression of BMSC markers in LV myocardium of dogs.

2. Experimental Model of Heart Failure in Dogs

LV dysfunction is produced by intracoronary microembolizations that result in loss of viable myocardium and the model has been fully characterized (3) and is well suited to determine the efficacy of experimental heart failure treatments (4–7). HF in this model is produced by multiple sequential intracoronary microembolizations that lead to loss of viable myocardium. The model manifests many of the sequelae of heart failure in humans including profound systolic and diastolic LV dysfunction, LV dilation and compensatory hypertrophy, increased LV filling pressures, increased systemic vascular resistance, and decreased cardiac output (3, 8). LV dysfunction in this model is accompanied by neurohumoral activation including sustained elevation of plasma norepinephrine, angiotensin-II, atrial natriuretic peptide, and endothelin-1 levels. As in patients, the model manifests, down-regulation of cardiac beta-adrenergic receptors (9), development of mild to moderate functional mitral regurgitation (10), LV shape changes (remodeling; increased chamber sphericity) (10, 11), development of chronic ventricular arrhythmias (12), third heart sound (13), exercise intolerance (14), ultrastructural abnormalities of residual viable cardiomyocytes (15), and accumulation of collagen in the cardiac interstitium. Of particular is the demonstration of spontaneous and progressive deterioration of LV function long after complete cessation of coronary microembolizations (3, 14). We have also shown that the acute hemodynamic response in dogs with HF to intravenous infusion of prototypical drugs, such as dobutamine, nitroprusside, enalaprilat, and digoxin are similar to responses observed in patients (5, 16).

3. Materials

3.1. Isolation of RNA from LV Myocardium of Dogs

1. 12 × 75 polystyrene tubes.
2. RNA Stat-60 (Tel-Test Inc. Friendswood, TX).
3. Polytron 7 mm homogenizer.
4. Chloroform (Ameresco).
5. Isopropanol (Ameresco).
6. Kim-wipes tissue.
7. 75% Ethyl alcohol.

3.2. Synthesis of cDNA from RNA (Reverse Transcription)

1. 0.2 ml thin walled autoclaved tubes.
2. Oligo dt primer (Invitrogen, Carlsbad, CA).
3. dATP, dTTP, dGTP, and dCTP (Invitrogen, Carlsbad, CA).
4. RnaseOut and superscript (Invitrogen, Carlsbad, CA).
5. Any thermocycler.

3.3. Amplification of cDNA by Real-Time Polymerase Chain Reaction

1. Real-Time polymerase chain reaction (PCR) is carried out in a 7500 fast real-time PCR system (Applied Biosystems).
2. RT real-time SYBR Green/ROX PCR mastermix (SuperArray).

3.4. Measurement of Protein Expression of Stem Cell Markers

1. Eppendorf tubes.
2. Homogenization buffer (HB) [50 mm Tris–HCl (pH 7.4), 0.5 mm Na$^+$ EDTA (pH 7.0), 0.3 M sucrose and protease inhibitors (0.8 mm benzamidine, 0.8 mg/l each aprotinin and leupeptin, and 0.4 μg/l antipain)].
3. Any sonicator.
4. Ice and water bath.
5. Preformed 12 well 4–20% Precise protein gel (Thermo Scientific, Rockford, IL).

6. SE-600 electrophoretic chamber (Hoeffer Scientific Instruments).

7. Transfer buffer (tris/glycine, 1× with 20% methanol).

8. Forceps or water gloves.

9. Precut filter paper (Whatman 3 mm).

10. Blocking buffer (5% nonfat dry milk in TBS-T, pH 7.6).

11. Primary antibody for Sca1 and cKit (Santa Cruz Biotechnology, Inc., Santa Cruz, CA).

12. ECL reagent.

3.5. Measurement of Circulating Sca1 Positive Bone Marrow Stem Cells in Blood

1. Heparinized tubes.

2. 15 ml tubes.

3. Phosphate buffer saline.

4. Lymphocyte isolation solution (Ficoll-Hypaque).

5. 17×100 mm culture tubes.

6. Fetal calf serum.

7. Hemocytometer and fluorescent microscope.

4. Methods

4.1. Isolation of RNA from LV Myocardium of Dogs

1. Weigh about 150 mg LV powder in 12×75 polystyrene tubes in such a way that the tissue remains in a frozen state. Use liquid nitrogen during weighing the samples.

2. Add 1.5 ml RNA Stat-60 (Tel-Test Inc. Friendswood, TX) into the tube containing tissue. Subsequently, vortex and incubate the tubes at room temperature for 5 min to allow the tissue to thaw.

3. Homogenize the tissue using a polytron 7 mm generator at setting 5 for 3×10 s with 5 s intermittent waiting.

4. Incubate the homogenate at room temperature for 5 min.

5. Add 0.4 ml chloroform (Ameresco) into the tubes and shake the tubes vigorously for 15 s.

6. Incubate the tubes at room temperature for 3 min and then centrifuge the tubes at $6,750 \times g$ for 15 min at 4°C.

7. Transfer the clear supernatant into a new polystyrene tube. Be careful not to disturb the middle or lower phases.

8. Add 0.9 ml isopropanol (Ameresco) and incubate the tube for 10 min at 37°C.

9. Centrifuge the tubes at $6,750 \times g$ for 10 min at 4°C.

10. Discard the supernatant carefully without disturbing the pellet and then wipe the rim of the tubes with sterilized Kim-wipes tissue.

11. Add 1 ml 75% ethyl alcohol. Vortex the tubes to loosen the RNA pellet. Do not try to resuspend the pellet.

12. Centrifuge the tubes at $6,750 \times g$ for 5 min at 4°C.

13. Discard the supernatant without loosing the pellet. Remember the pellet is very loose at this step.

14. Keep the tubes in inverted order over sterilized Kim-wipes tissue paper and absorb the excess liquid from the walls of inside of the tubes by using autoclaved cotton tipped applicators.

15. Let the tube air dry for 15 min at room temperature. At this point, do not use speed vacuum.

16. Add 100 μl of deionized distilled water. Vortex and incubate the tubes at 55–60°C to dissolve RNA pellet.

17. Dilute an aliquot of 10 μl to 1 ml in deionized distilled water and take optical density at 260 and 280 nm in a spectrophotometer. If the ratio of OD260/OD280 for RNA is close to 1.9–2.0, this shows that RNA is intact and of good quality.

4.2. Synthesis of cDNA from RNA (Reverse Transcription)

1. Total RNA is diluted to a 0.1 mg/ml concentration in deionized distilled water and denatured at 95°C for 5 min, followed by rapid cooling in an ice bath.

2. In 0.2 ml thin-walled autoclaved tubes, add 50 μl denatured RNA obtained from the step above and add 50 μl mix containing 9 μl distilled water, 1 μl Oligo dt primer, 5 μl dNTP mix (dATP, dTTP, dGTP, and dCTP), 20 μl first strand buffer, 10 μl 0.1 M DTT, and 2.5 μl each of RnaseOut and Superscript.

3. The tubes containing the samples are incubated in a Bio-Rad icycler at 42°C for 1 h, after which the reaction is terminated at 95°C for 5 min.

4.3. Amplification of cDNA by Real-Time Polymerase Chain Reaction

1. Each PCR reaction is carried out in a 96 well plate specifically made for 7500 fast real-time PCR system with an assay volume of 20 μl, including 1 μl of cDNA, 0.5 μl (20 pmole) of forward and 0.5 μl (20 pmole) of reverse gene specific primers for stem cell markers Sca1, cKit and a house keeping gene glyceraldehyde-3 phosphate dehydrogenase (GAPDH), 10 μl of RT real-time SYBR Green/ROX PCR mastermix (SuperArray) and 8 μl of deionized distilled water. The PCR mixture is amplified by a 7500 fast real-time PCR system (Applied Biosystems).

2. A hot start is given at 95°C for 10 min, and amplification is performed for 50 cycles according to the following program: denaturing at 94°C for 15 s, annealing at 60°C for 1 min, and later a dissociation cycle is carried out at 95°C for 15 s followed by 60°C for 1 min and then 95°C for 15 s. The dissociation curve is very important because it lets us know about any dimers or nonspecific bands which are formed during the PCR.

3. The PCR cycler will automatically give the Ct values for all samples. For quantification of PCR reaction products, Ct values of stem cell markers are subtracted from GAPDH values and the value obtained is referred to a $\Delta\Delta$Ct. Then the final value of mRNA expression is calculated by 2 raised to the power of $-\Delta\Delta$Ct.

4.4. Preparation of SDS-Extract from LV Myocardium of Dogs

Measurement of protein expression is carried out as previously described (17–21).

1. 25 mg of cardiac powder is weighed in a prechilled (liquid nitrogen) Eppendorf tube and stored in –70°C freezer.

2. After the samples are weighed, take out all the samples from the freezer, keep them in ice-bath, and then immediately add 0.6 ml HB into all tubes.

3. Let the frozen tissue thaw in HB. Within 8 min, vortex all the samples four times at an interval of 2 min.

4. Homogenize the samples by a sonicator at a setting of 25/40 for 3×20 s. During homogenization, samples must be in the ice-bath.

5. Add 0.6 ml 10% SDS to all tubes and then vortex.

6. Incubate all the samples simultaneously in a boiling water bath for 10 min. During boiling, vortex samples two times at an interval of 5 min.

7. Cool the samples simultaneously at room temperature. It takes about 15–20 min.

8. Centrifuge the samples at 9,000×g at room temperature for 20 min.

9. Carefully pipette out 900 μl clear supernatant into another prelabeled tube. Make sure that the supernatant does not mix with the pellet while pipetting out the supernatant.

10. Protein assay is carried out by using the BioRad DC protein assay kit.

4.5. SDS Page

1. 25 μg of protein for each sample and a marker is loaded on a preformed 12 well 4–20% Precise protein gel (Thermo Scientific, Rockford, IL). The gel is run at 120 V for 2 h till

molecular marker runs off bottom of the gel in a SE-600 electrophoretic chamber (Hoeffer Scientific Instruments) containing running buffer (tris glycine/SDS, 1×).

4.6. Transfer of Protein Bands from Gel to Nitrocellulose Membrane

1. Following electrophoresis, rinse the gels in 50 ml transfer buffer (tris/glycine, 1× with 20% methanol), once for 5 min and second time for 10 min to facilitate the removal of electrophoresis buffer salts and detergents which are responsible for heat generation during transfer.

2. Cut the membrane to the dimension of the gel and label it with a soft pencil to identify the gel and the orientation of the membrane. Wet the membrane by slowly sliding it at a 45° angle into 50 ml of transfer buffer and allow it to soak for 10–15 min. To avoid membrane contamination, always use forceps or water gloves when handling the membrane.

3. Completely saturate the precut filter paper (Whatman 3 mm) and fiber pads by soaking them in transfer buffer.

4.7. Assembly of the Gel and Membrane

1. Place the opened gel holder in a shallow vessel so that the black panel is flat on the bottom of the vessel.

2. Place a presoaked fiber pad on the black panel of the cassette. When assembling the fiber pads, filter paper, gel, and membrane, be sure to center all components.

3. Place a piece of the saturated filter paper on top of the fiber pad. Saturate the surface of the filter paper by 2–3 ml of transfer buffer. Place the gel on top of the paper. Align the gel in the center of the cassette. Make sure that no air bubbles are trapped between the gel and the filter paper.

4. Flood the surface of the gel with transfer buffer and lower the prewetted membrane on top of the gel. This is best done by holding the membrane at opposite ends so that the center will contact the gel first. Then gradually lower the ends. Next, roll a glass pipette or test tube over the top of the membrane to exclude all air bubbles from the area between the gel and membrane. Note: light pressure should be applied until a nearly adhesive contact is made between the membrane and gel.

5. Flood the surface of the membrane with transfer buffer. Complete the sandwich by placing a piece of saturated filter paper on top of the membrane and placing a saturated fiber pad on top of the filter paper. Now close the cassette. Hold it firmly so the sandwich will not move and secure the latch.

6. Place the gel holder in the buffer tank so that the black panel of the holder is facing the black cathode electrode panel. Note: gel is facing cathode and membrane anode.

7. Prepare the bio-Ice cooling unit in advance, by filling it with de-ionized, distilled water and storing it in the freezer. Install the frozen cooling unit in the buffer chamber, next to the electrode, a few minutes before starting the transfer.

8. When the gels to be transferred are in place, set the buffer tank on top of a magnetic stirrer. Fill the tank with cold transfer buffer to just above the level of the top row on the gel holder cassette. Turn on the magnetic stirrer, and put the lid in place.

9. Turn on the power supply to initiate transfer. Overnight: 30 V; Rapid 2 h: 100 V.

10. To check transfer of proteins, stain the membrane with 0.01% naphthol blue–black dye for 1 min and immediately destain in deionized water with several changes till background becomes clear. This staining is optional and is done to check for proper transfer from gel to membrane.

4.8. Blocking and Incubation of Membrane with Antibodies

1. The nitrocellulose membrane is then incubated in blocking buffer (5% nonfat dry milk in TBS-T, pH 7.6) for 1 h at room temperature on a shaker. The blocking buffer is discarded and the membrane is rinsed with TBS-T.

2. Primary antibody for Sca1 and cKit (Santa Cruz Biotechnology, Inc., Santa Cruz, CA) is diluted to 1:500 in antibody buffer (1% non fat dry milk in TBS-T, Ph 7.6) and the membrane is incubated overnight: in cold room on a shaker or for 2 h at room temperature.

3. The primary antibody is removed and the membrane is washed three times for 10 min each with 20 ml TBS-T.

4. The secondary antibody is prepared at a concentration of 1:5,000 in antibody buffer and the membrane is incubated for 2 h at room temperature.

5. The secondary antibody is discarded and the membrane is washed three times for 15 min each with 20 ml TBS-T.

6. Two-and-a-half milliliter aliquots of each portion of the ECL reagent are mixed together and added to the blot after discarding the last wash. The membrane is shaken by hand for 2–4 min.

7. The blot is removed from ECL and excess ECL is removed by Kim-wipes before placing the blot between an acetate sheet protector.

8. Pictures of the blot are immediately taken in a Molecular ChemiDoc XRS chemiluminescent documentation system (Bio-Rad).

4.9. Measurement of Circulating Sca1 Positive Bone Marrow Stem Cells in Blood

Circulating, Sca1-positive BMSCs were isolated from whole blood as previously described (17, 22). Sca1-positive BMSC cells include mesenchymal, multipotent adult progenitor, and Hoechst side population cells (17, 22, 23).

1. Venous blood is collected from dogs in two heparinized tubes.

2. Transfer blood from both collection tubes to one 15 ml tube.

3. Add 7 ml of phosphate buffer saline (PBS) to blood and mix well. Total volume will be 14 ml.

4. Add 4.6 ml lymphocyte isolation solution (Ficoll-Hypaque) to two culture tubes (17×100 mm). Now very gently add ~7 ml of diluted blood on top of Ficoll-Hypaque solution. Prepare two tubes for each sample.

5. Centrifuge the tubes in swinging bucket rotor at $400 \times g$ for 30–40 min at room temperature.

6. After centrifugation, lymphocytes along with BMSCs are visible as a white ring in the center of tube. Aspirate off the solution from top of lymphocyte ring and carefully transfer the white ring from both tubes to fresh 17×100 mm culture tube.

7. Add 4.5 ml PBS to isolated lymphocytes and centrifuge at $260 \times g$ for 10 min and discard the supernatant.

8. Isolated cells are blocked with 10% normal blocking serum (1 ml fetal calf serum and 9 ml PBS) for 30 min to suppress nonspecific binding. Then cells are washed with PBS.

9. Cells are kept in the cold room to incubate with the primary antibody for 2 h (Santa Cruz Biotechnology, Inc., Santa Cruz, CA) at a concentration of 1:50 in 1.5% blocking serum.

10. Cells are once again washed with PBS and incubated with secondary antibody containing fluorescein isothiocynate (FITC) for 1 h in dark and on ice. Cells are washed, centrifuged and then resuspended in PBS.

11. Sca1 positive BMSCs are now fluorescent and are counted using a hemocytometer coupled to fluorescent microscope.

12. The hemocytometer consists of nine 1-mm squares divided into smaller squares. One of the 1-mm squares represents a volume of 0.1 mm^3 or 10^{-4} ml. Using the 10× objective, the total number of cells and fluorescent cells in a 1-mm square area are counted. If there are fewer than 100 cells in a square mm, two or more 1-mm square areas are counted and the results averaged.

5. Notes

1. Be very careful while isolating RNA and do not touch the middle or lower phases while taking out the supernatant.

2. Once the RNA pellet is in ethyl alcohol, do not try to resuspend the pellet.

3. Never use speed vacuum to dry the tubes.

4. The dissociation curve in RT-PCR is very important because it tells us about any dimers or nonspecific bands that are formed during the PCR.

5. Make sure that the supernatant does not mix with the pellet while pipetting out the supernatant when making samples for protein measurement.

6. To avoid membrane contamination, always use forceps or water gloves when handling the nitrocellulose membrane.

7. It is very important to make sure that no air bubbles are trapped between the gel and the filter paper and membrane.

8. Gel should always be facing cathode and membrane anode or else transfer will not take place.

9. Be very careful and do not disturb the white ring containing BMSCs obtained after Ficoll-Hypaque centrifugation.

References

1. Hasenfuss, G. (1998) Animal models of human cardiovascular disease, heart failure and hypertrophy, *Cardiovasc Res 39*, 60–76.

2. Rastogi, S., Gupta, R. C., Mishra, S., Morita, H., and Sabbah, H. N. (2002) Long term therapy with Acorn cardiac support device normalizes gene expression of alfa and beta myosin heavy chain in dogs with chronic heart failure, *Circulation 106*(19), II–384 *Abstract*.

3. Sabbah, H. N., Stein, P. D., Kono, T., Gheorghiade, M., Levine, T. B., Jafri, S., Hawkins, E. T., and Goldstein, S. (1991) A canine model of chronic heart failure produced by multiple sequential coronary microembolizations, *Am J Physiol 260*, H1379–H1384.

4. Morita, H., Suzuki, G., Mishima, T., Chaudhry, P. A., Anagnostopoulos, P. V., Tanhehco, E. J., Sharov, V. G., Goldstein, S., and Sabbah, H. N. (2002) Effects of long-term monotherapy with metoprolol CR/XL on the progression of left ventricular dysfunction and remodeling in dogs with chronic heart failure, *Cardiovasc Drugs Ther 16*, 443–449.

5. Sabbah, H. N., Shimoyama, H., Kono, T., Gupta, R. C., Sharov, V. G., Scicli, G., Levine, T. B., and Goldstein, S. (1994) Effects of long-term monotherapy with enalapril, metoprolol, and digoxin on the progression of left ventricular dysfunction and dilation in dogs with reduced ejection fraction, *Circulation 89*, 2852–2859.

6. Suzuki, G., Morita, H., Mishima, T., Sharov, V. G., Todor, A., Tanhehco, E. J., Rudolph, A. E., McMahon, E. G., Goldstein, S., and Sabbah, H. N. (2002) Effects of long-term monotherapy with eplerenone, a novel aldosterone blocker, on progression of left ventricular dysfunction and remodeling in dogs with heart failure, *Circulation 106*, 2967–2972.

7. Tanimura, M., Sharov, V. G., Shimoyama, H., Mishima, T., Levine, T. B., Goldstein, S., and Sabbah, H. N. (1999) Effects of AT1-receptor blockade on progression of left ventricular dysfunction in dogs with heart failure, *Am J Physiol 276*, H1385–H1392.

8. Kono, T., Sabbah, H. N., Rosman, H., Alam, M., Stein, P. D., and Goldstein, S. (1992)

Left atrial contribution to ventricular filling during the course of evolving heart failure, *Circulation 86*, 1317–1322.

9. Gengo, P. J., Sabbah, H. N., Steffen, R. P., Sharpe, J. K., Kono, T., Stein, P. D., and Goldstein, S. (1992) Myocardial beta adrenoceptor and voltage sensitive calcium channel changes in a canine model of chronic heart failure, *J Mol Cell Cardiol 24*, 1361–1369.

10. Sabbah, H. N., Kono, T., Rosman, H., Jafri, S., Stein, P. D., and Goldstein, S. (1992) Left ventricular shape: a factor in the etiology of functional mitral regurgitation in heart failure, *Am Heart J 123*, 961–966.

11. Sabbah, H. N., Kono, T., Stein, P. D., Mancini, G. B., and Goldstein, S. (1992) Left ventricular shape changes during the course of evolving heart failure, *Am J Physiol 263*, H266–H270.

12. Sabbah, H. N., Goldberg, A. D., Schoels, W., Kono, T., Webb, C., Brachmann, J., and Goldstein, S. (1992) Spontaneous and inducible ventricular arrhythmias in a canine model of chronic heart failure: relation to haemodynamics and sympathoadrenergic activation, *Eur Heart J 13*, 1562–1572.

13. Kono, T., Rosman, H., Alam, M., Stein, P. D., and Sabbah, H. N. (1993) Hemodynamic correlates of the third heart sound during the evolution of chronic heart failure, *J Am Coll Cardiol 21*, 419–423.

14. Sabbah, H. N., Hansen-Smith, F., Sharov, V. G., Kono, T., Lesch, M., Gengo, P. J., Steffen, R. P., Levine, T. B., and Goldstein, S. (1993) Decreased proportion of type I myofibers in skeletal muscle of dogs with chronic heart failure, *Circulation 87*, 1729–1737.

15. Sabbah, H. N., Sharov, V., Riddle, J. M., Kono, T., Lesch, M., and Goldstein, S. (1992) Mitochondrial abnormalities in myocardium of dogs with chronic heart failure, *J Mol Cell Cardiol 24*, 1333–1347.

16. Sabbah, H. N., Levine, T. B., Gheorghiade, M., Kono, T., and Goldstein, S. (1993) Hemodynamic response of a canine model of chronic heart failure to intravenous dobutamine,
nitroprusside, enalaprilat, and digoxin, *Cardiovasc Drugs Ther 7*, 349–356.

17. Rastogi, S., Imai, M., Sharov, V. G., Mishra, S., and Sabbah, H. N. (2008) Darbepoetin alfa prevents progressive left ventricular dysfunction and remodeling in non-anemic dogs with heart failure, *Am J Physiol Heart Circ Physiol 295*, H2475–H2482.

18. Rastogi, S., Mishra, S., Zaca, V., Alesh, I., Gupta, R. C., Goldstein, S., and Sabbah, H. N. (2007) Effect of long-term monotherapy with the aldosterone receptor blocker eplerenone on cytoskeletal proteins and matrix metalloproteinases in dogs with heart failure, *Cardiovasc Drugs Ther 21*, 415–422.

19. Rastogi, S., Mishra, S., Zaca, V., Mika, Y., Rousso, B., and Sabbah, H. N. (2008) Effects of chronic therapy with cardiac contractility modulation electrical signals on cytoskeletal proteins and matrix metalloproteinases in dogs with heart failure, *Cardiology 110*, 230–237.

20. Rastogi, S., Sentex, E., Elimban, V., Dhalla, N. S., and Netticadan, T. (2003) Elevated levels of protein phosphatase 1 and phosphatase 2A may contribute to cardiac dysfunction in diabetes, *Biochim Biophys Acta 1638*, 273–277.

21. Rastogi, S., Sharov, V. G., Mishra, S., Gupta, R. C., Blackburn, B., Belardinelli, L., Stanley, W. C., and Sabbah, H. N. (2008) Ranolazine combined with enalapril or metoprolol prevents progressive LV dysfunction and remodeling in dogs with moderate heart failure, *Am J Physiol Heart Circ Physiol 295*, H2149–H2155.

22. Zaca, V., Rastogi, S., Imai, M., Wang, M., Sharov, V. G., Jiang, A., Goldstein, S., and Sabbah, H. N. (2007) Chronic monotherapy with rosuvastatin prevents progressive left ventricular dysfunction and remodeling in dogs with heart failure, *J Am Coll Cardiol 50*, 551–557.

23. Kotton, D. N., Summer, R. S., Sun, X., Ma, B. Y., and Fine, A. (2003) Stem cell antigen-1 expression in the pulmonary vascular endothelium, *Am J Physiol Lung Cell Mol Physiol 284*, L990–L996.

Part III

Histological Assessment of Myocardial Regeneration

Chapter 9

Histopathologic Assessment of Myocardial Regeneration

Naima Carter-Monroe, Elena Ladich, Renu Virmani, and Frank D. Kolodgie

Abstract

Cardiac regeneration in the form of cell-based therapy offers hope of becoming the breakthrough technology that transforms the state of cardiac medicine. Before attempting to develop the techniques to assess the effectiveness of myocardial regeneration in humans, researchers must have at least a basic understanding of the human heart in its embryonic, normal, and diseased states. To this end, we provide an overview of the histology of the heart, including the current theories on normal embryogenesis and the histology of normal and ischemic myocardium as visualized by pathologists. Knowledge of the cellular constituents, including the controversial existence of resident cardiac stem and/or progenitor cells, and their actions and interactions in the normal state and under the conditions of myocardial ischemia is also crucial before embarking on the quest for cardiac regeneration. Despite widespread optimism in the success of cell-based therapy, inherent difficulties remain in the identification of effective cell populations proposed for cell-based therapy in the human heart.

Key words: Acute myocardial infarction (AMI), Cardiomyopathy, Cardiomyocytes, Myocardial regeneration, Myofibroblast, Ventricular remodeling, Cardiac stem cell, Cardiac progenitor cell, c-Kit, Sca-1, MDR1

1. Introduction

Cardiovascular disease is responsible for 37% of deaths in the United States (1). Congestive heart failure constitutes a major public health problem for individuals above 50 years old where the incidence increases progressively with age. In the United States alone, 4.9 million Americans have been diagnosed with heart failure (2). The rate of hospital discharges for heart failure has increased by 155% in the last 20 years, and heart failure is the most frequent cause of hospitalization in persons aged 65 years or older. These statistical observations highlight the relationship of

Randall J. Lee (ed.), *Stem Cells for Myocardial Regeneration: Methods and Protocols*, Methods in Molecular Biology, vol. 660, DOI 10.1007/978-1-60761-705-1_9, © Springer Science+Business Media, LLC 2010

the aging patient with an increased susceptibility to incur ischemic damage since there is diminished functional and adaptive reserve capacity and lack of practical ability for regeneration and repair (3, 4).

Ischemic heart disease remains the major etiology of heart failure, which has experienced an increase from 29 to 52% over recent decades (5). Other factors responsible include hypertension, valvular heart disease, atrial fibrillation, alcohol, and cardiomyopathies. The treatment options however, remain limited to pharmacotherapy and transplantation where there are urgent needs to pursue strategies that will restore heart function regardless of the underlying etiology. Of late, recent discoveries in developmental and adult-myocardial-cell biology have optimistically led to a more dynamic picture of myocardial cell biology and homeostasis, thereby creating newer opportunities for regenerative cardiovascular medicine.

Cell therapy, predicated on the exogenous delivery of pluri- or multipotent stem cells (or stimulation of endogenous "resident" myocyte precursor cells that potentially exist within the adult heart) is developing into a viable strategy for the treatment of cardiomyopathy. Further inroads, however, have to be made considering well over 15 years of research focused on repopulating the infarcted heart with exogenous cells has provided little success. Previous attempts at restoring myocardial function with cell-based therapy have included skeletal myoblasts, fetal or neonatal cardiomyocytes, hematopoietic stem cells, mesenchymal stem cells, endothelial cells, resident cardiac progenitor cells, and embryonic stem cells (ECSs). Clearly most studies, irrespective of the transplanted cell type, have shown promising effects consistent with functional improvement presumably based on the successful formation of electrically coupled myocardium with the host heart or enhanced development of collaterals. The fact that such diverse cell preparations can enhance cardiac repair, when many cannot even generate new muscle tissue, suggests that multiple mechanisms may be involved, mechanisms at this time, we clearly do not understand.

2. The Normal Myocardium

2.1. Embryogenesis

Embryonic development in the mouse suggests that myocardial development begins prior to embryonic day 9.5. At this time, the heart structure forms as a thin-walled linear tube structure with morphologically indistinct, but molecularly specified atrial and ventricular chambers. Along the periphery of this tube, myocytes proliferate in single or double layers of cells that form the compact region (6). Because the primitive heart is without arterial blood supply, its continued growth is limited by diffusion of nutrients from its lumen and the organ can only continue to grow by trabeculation. Trabeculation is the process by which the ventricles begin to thicken while the myocytes along the inner wall become organized into fingerlike projections (called trabeculae). These trabeculae are organized in several intertwined helices (6). Studies have shown differences in the trabeculation pattern between the ventricles of different species (6–8). It has also been shown that retinoic acid produced by the epicardium is a critical regulator of cardiac growth. At birth, the myocardium is only a fraction of the thickness it will become during adulthood where it continues to increase at least fourfold in postnatal life as a result of myocyte hypertrophy rather than cell division. The latter point becomes important as we discuss the possibility of the existence of a reserve of cardiac stem cells with the ability to undergo cell division.

Current convention divides the progenitor cells of the various cellular components of the heart into primary (PHF) and secondary heart field lineages (SHF) (9–12). In brief, the primary heart field develops from the anterior lateral mesoderm to form the early embryologic stage of the cardiac crescent, which progresses to the heart tube and ultimately the left ventricle. Derived from migrating cells arising from the pharyngeal and splanchnic mesoderm, the SHF progenitor cells form the right ventricle, outflow tract, and part of the inflow tract. Current lineage-tracing experiments have been only partially successful at identifying specific molecular markers for the PHF progenitor cells. Both the cardiac specific enhancer NKx2.5 and the cell surface marker Flk1 have been shown to identify a mix of PHF and SHF progenitors, whereas transcription factor Islet 1 [Isl1] appears to be specific for SHF progenitors only (9, 12–14).

As the coronary circulation becomes functional, the trabeculae become compressed and form most of the compact layer of the heart (15). Failure of compaction can manifest as significant cardiac dysfunction as illustrated in the human disorder of isolated ventricular noncompaction (16). Several studies have revealed that the trabeculae are the first portion of the heart to undergo electrical activation, and activation propagates radially toward the outside of the heart tube (17, 18). According to work by Sedmara, electrical activation in the early heart follows the direction of blood flow, and thus explains a transition from base-to-apex to mature apex-to-base propagation of electrical impulses (6).

2.2. Histology of the Normal Functioning Myocardium

In the simplest anatomic terms, the functional unit of normal myocardium is the cardiac myocyte, and the mass of myocyte bundles, arranged in spiral and circumferential orientation, forms the myocardium, which is the muscular portion of the heart. Cardiac myocytes are a specific subset of muscle cell that histologically appear as elongated and tapered with a low nuclear to cytoplasmic ratio (N/C ratio), centrally placed nuclei, and

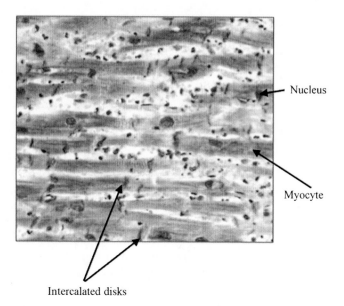

Intercalated disks

Fig. 1. Architecture of normal cardiac myocytes – light microscopic image of paraffin embedded myocardium (Desmin stain) illustrating myocytes with intercalated disks (darker staining end-to-end junctions) and faint, slightly darker staining cross striations. Also note interstitial tissue (*light blue staining*) interspersed between myocytes.

end-to-end cell junctions that are visualized on H&E stains as slightly darker stained intercalated disks (see Fig. 1). Because of these end-to-end junctions, cardiac myocytes exist as a syncytium, which allows them to act as larger functional units – otherwise known as myofibrils (19). Within the intercalated disks are specialized intracellular junctions, such as gap junctions, which allow for the electrical and mechanical coupling of adjoining cells.

Individual myocytes are composed of: (1) a nucleus; (2) the sarcolemma or cell membrane with T-tubules to conduct electrical impulses; (3) the sarcoplasmic reticulum, which serves as a calcium reservoir; (4) multiple mitochondria; and (5) the contractile elements. We visualize the contractile elements under the microscope as striations. Upon ultrastructural examination, we can see that the striations are composed of parallel arrangements of myofilaments. Myofilaments are parallel arrangements of sarcomeres, and it is the sarcomere that acts as the functional intracellular contractile element of the cardiac myocyte. Sarcomeres are composed of (1) thick filaments or myosin and (2) thin filaments composed of actin. It is the overlapping arrangement of and ability of the thick and filaments to "slide" over each other, thus shortening the cell that enables myocyte contraction. Attached to the myofilaments are the regulatory proteins of the troponin family and tropomyosin. In addition, cardiac myofilaments have many more mitochondria associated with them (at 23% cell volume) than in other muscle cell types (2% cell volume) (19).

Given the function of the cardiac myocyte, it is not surprising that although cardiac myocytes only account for 25% of the total number of cells in the heart, cardiac myocytes comprise 90% of the total heart mass (19). Cardiac myocytes populate both the atria and ventricles, with atrial cardiac myocytes usually being the smaller of the two. In general, ventricular myocytes are longer and narrower (average length between 60 and 140 μm) with prominent end-to-end junctions. By comparison, atrial cardiac myocytes are shorter, at a length of ~20 μm, and possess both end-to-end and side-to-side junctions. Atrial cardiac myocytes may also be differentiated from their ventricular counterparts by the presence of electron dense granules (specific atrial granules) in their cytoplasm, which stores heart-specific molecules such as atrial natriuretic peptide (ANP). Clearly visible in other mammalian hearts, but less apparent in human hearts, are the long and broad cardiac purkinje cells. Purkinje cells appear pale on routine H&E stains due to fewer sarcomeres in their cytoplasm as compared to both atrial and ventricular cardiac myocytes.

In addition to the cardiac myocyte, there are interspersed vessels ranging from capillaries to small intracardiac arteries (composed of cardiac endothelial cells), myofibroblasts, and usually negligible amounts of collagen and inflammatory cells. Capillaries are abundant in the myocardium and by our observations there are four to five capillaries for every myocyte. In the normal myocardium, capillary density in the epicardium has been reported at 2,439 capillaries/mm^2, which is 21% higher than the density in the endocardium (2,014 capillaries/mm^2) (20). Later in the discussion, we will focus on the myofibroblast and its importance in the infarct scar and ventricular remodeling.

3. Myocardial Ischemia

3.1. Factors Contributing to the Healing Process After AMI

Slower healing is observed in larger infarcts as compared to smaller infarcts. This phenomenon is due to propensity of infarcts to heal faster as the surface/volume ratio of the infarct increases. As an infarct enlarges, there is an inversely proportional decrease in the surface/volume ratio that results in slower healing (21). The relationship of the size of injury to the rate of healing is further exemplified by the fact that reperfused infarcts heal at a faster rate than their nonreperfused counterparts.

3.2. Histopathology (Light Microscopic Appearance) of the Nonreperfused Infarct

The earliest of pathologic changes in 12–24 h nonreperfused infarcts, by light microscopy, is the presence of hypereosinophilic myocytes destined for necrosis. The injured myocytes are elongated with poorly visualized cell striations and cell borders. Degenerating (karyorrhectic) nuclei are often seen. Early hypereosinophilic myocytes may still show nuclei and striations with some adjacent inconspicuous interstitial edema. Acute inflammatory cells (neutrophils) begin to aggregate at the border of the infarct 24 h post coronary artery occlusion, and markedly increase between 24 and 48 h with the onset of coagulation necrosis. There is progressive loss of myocyte nuclei and cell striations starting from the center of the infarct continuing 3–5 days post occlusion. During this same time interval, macrophages and myofibroblasts begin to move toward the infarct border zone (22).

Cardiomyocyte degradation occurs at approximately 8 days post infarction where infiltrating macrophages and angiogenic vessels facilitate the complete removal of the necrotic tissue. The composition of the infarct at this time consists primarily of granulation tissue characterized by chronic inflammatory cells (mostly macrophages, lymphocytes, and plasma cells), loose fibrous tissue, myofibroblasts, and neoangiogenic vessels. Along the infarct border zone (defined as the interface between the infarct and adjacent "normal" myocardium), bizarre, multinucleated myocytes are commonly present (see Fig. 2). At this stage, there is a

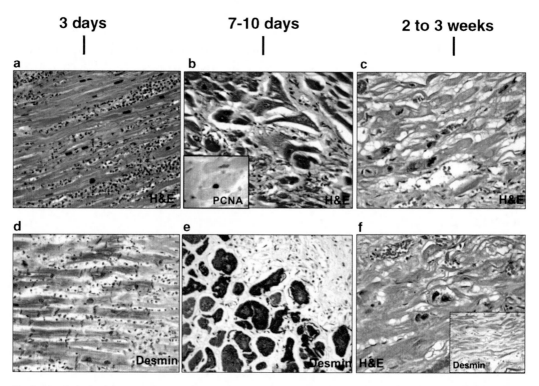

3 days   **7-10 days**   **2 to 3 weeks**

Fig. 2. Morphologic changes at 3, 7–10 days, and 2–3 weeks following myocardial infarction in man. *Top row* (**a–c**) shows H&E stained sections with marked acute inflammatory infiltrate in (**a**), early granulation tissue and adjoining bizarre and multinucleate but viable myocytes in (**b**), and (**c**) shows border area with multinucleated cells at 2–3 weeks. The myocytes shown in (**b**) are PCNA positive (insert in **b**). (**d**) and (**e**) are adjoining sections to (**a**) and (**b**) and are stained with antidesmin stains; viable myocytes are stained by **DAB** (*brown*). (**f**) is a different area than shown in (**c**) and shows multinucleated myocytes, which stain positive with antidesmin antibodies.

notable dampening of the acute inflammatory response. The infarct continues to heal and by 2 weeks, there is replacement of necrotic tissue by collagen principally secreted by myofibroblasts. The infarct becomes increasingly enriched in collagen by 6 weeks, which is interspersed with islands of intact myocytes more frequently observed in reperfused than nonreperfused infarcts (see Fig. 3). Infarct healing is generally complete by 4–8 weeks although persistent "mummified" myocytes may be present in center of infarct; this is the main reason why pathologic dating of ischemic events should be restricted to infarct border zones (22).

3.3. Histopathology (Light Microscopic Appearance) of the Reperfused Infarct

Reperfused myocardial infarcts 4–6 h after a patient experiences chest pain and/or ECG changes manifest typically show hemorrhage and contraction band necrosis superimposed in the area of injured myocardium. The occurrence of hemorrhage with neutrophil accumulation takes place within a few hours (as compared to nonreperfused MIs at 24–48 h) where fewer neutrophils (as compared to the nonreperfused infarcts) are found scattered throughout the infarct region. Macrophage migration into the infarct bed occurs at 2–3 days and even though fibroblasts are present at same time, compared to nonreperfused MI, the healing process is accelerated. Viable myocytes in border regions are

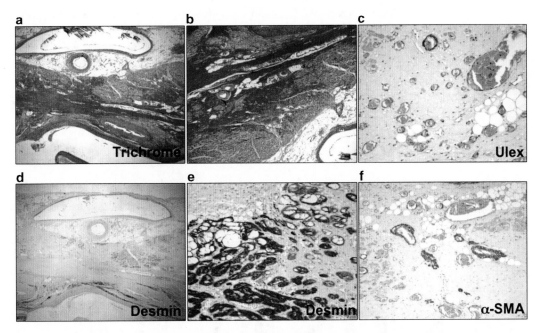

Fig. 3. Images from ≥2 month old healed myocardial infarct; (**a**) and (**b**) sections, stained by Masson's trichrome show the area of healed myocardial infarction; note the collagenized areas of the scar are *blue* while the viable myocytes are *red*. (**c**) The scarred area is stained using Ulex europaeus lectin to identify capillaries, arterioles, muscular arteries, and venules. (**d** and **e**) are desmin-stained sections showing focal positivity indicating only focal collections of viable myocytes (*brown stain*). The myocytes close to the endocardium in (**e**) are vacuolated but still viable. (**f**) is a myocardial section stained with ∝-smooth muscle actin showing smooth muscle cells around established vessels.

found in greater numbers in reperfused infarcts where their numbers are increased with shorter times of coronary occlusion. Subendocardial infarcts in the reperfused heart may be healed in as little as 2–3 weeks (22).

3.4. Composition of the Myocardial Scar

It is now appreciated that the infarct scar is indeed living tissue composed of active cell populations, vessels, and nerve tissue. The infiltration of myofibroblast, as described above, is one of the earliest events in scar formation where the appearance of these cells persists for years. Using the mouse infarct model, Virag and Murray demonstrated that myofibroblasts proliferate more rapidly than endothelial cells ($15.4 \pm 1.1\%$ vs. $2.9 \pm 0.5\%$, respectively) at 4 days after coronary occlusion, which sharply declines by 1 week to ($4.1 \pm 0.6\%$ and $0.7 \pm 0.1\%$) and is less than 0.5% after 2 weeks (23). The individual myofibroblast function to produce fibrillar collagens (I and III) providing the structural scaffolding of the scar in addition to producing soluble factors such as angiotensin II and transforming growth factor-β1, which regulate collagen (type I) turnover. Since these factors are soluble, collagen can accumulate in regions distant from the infarct; this unfortunate property is considered the underlying mechanism of ischemic cardiomyopathy (24, 25). Moreover, myofibroblasts are capable of binding to the extracellular matrix by utilizing α-SM actin-positive microfilaments thereby contributing to the contractile behavior of the scar. Innervation by postganglionic nerve fibers and neovascularization offers a supporting role while accompanying the myofibroblasts in the scar (24). By our estimates, capillary density within the region of the scar is approximately 36 capillaries/mm^2 representing a relatively large percentage of infarct area occupied by vessels at 7.95% compared to 1.98% of area occupied by capillaries in normal myocardium ($p = 0.05$) (see Fig. 4) (26).

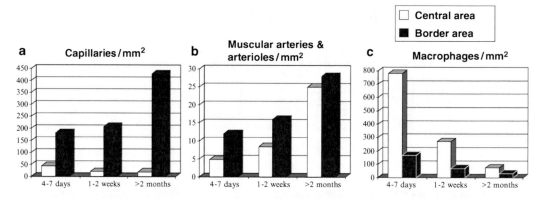

Fig. 4. A series of bar graphs showing the number of capillaries (**a**), muscular arteries and arterioles (**b**), and macrophage infiltration (**c**) in the border and central regions of human infarcts of age 4–7 days, 1–2 weeks, and ≥2 months.

3.5. Factors in Ventricular Remodeling

Ventricular remodeling has been implicated in left ventricular dilatation and dysfunction following an acute myocardial infarction (AMI) (27). In particular, the ability to limit or reverse the amount extracellular matrix proteins deposited in the infarct bed, and thus limiting fibrosis, may be of utility in slowing the remodeling process. The obvious targets for antifibrotic therapy are collagens given that 85 and 11% of the total deposited collagen consist of types (I and III), respectively. Collagen type I is composed of thick fibers that determine the tensile strength and resistance to stretch and deformation of the infarct while collagen III determines the resilience of the infarct (28, 29). Other extracellular matrix proteins of significance within the infarct scar include elastin, glycosoaminoglycans, integrins, fibronectin, and laminin. Matrix metalloproteinases (MMPs) function to degrade the extracellular matrix and in murine models the increased activity of MMPs leads to infarct rupture. In the study by van den Borne et al., areas with infarct rupture sites show increased MMP-8 and MMP-9 together with inflammatory cells (30). Tissue inhibitors of MMPs (TIMPS) also localize to the infarct site to counterbalance the degradation of the ECM by MMPs (27).

Additional factors with direct influence on the healing of an infarct include age, gender (presence of estrogen), nitric oxide, the renin-angiotensisn-aldosterone system, cytokines, leukocytes, chemokines, and drugs. Many of these factors have both positive and negative effects on infarct repair and ventricular modeling, highlighting the complexity of the process. For example, although inflammatory infiltration of the infarcted region is essential for the initiation of the healing, as cited above inflammatory cells can secrete destructive levels of MMPs. This mechanism may apply to older patients dying of myocardial infarction where deposited extracellular matrix is eventually degraded accompanied by increased neovascularization (31).

4. Cardiomyopathy: Histologic Characteristics as Compared to Myocardial Infarction

Given the breadth of entities known as the cardiomyopathies, the discussion will focus on only those disorders potentially amenable to cardiac regeneration therapy. These would include (idiopathic) dilated cardiomyopathy (IDCM), hypertrophic cardiomyopathy, arrhythmogenic right ventricular dysplasia, restrictive cardiomyopathies including the idiopathic, endocardial fibrosis and cardiac amyloidosis variants, and miscellaneous infantile and hereditary cardiomyopathies. Common to all these disease entities are the nonspecific findings of variable amounts of myocyte loss with fibrosis and myocyte hypertrophy. Hypertrophic cardiomyopathy is mainly characterized by myofiber disarray, although varying levels of interstitial and replacement fibrosis may be seen in ventricular septum. In both the dilated phase of hypertrophic cardiomyopathy and IDCM, extensive scarring similar to an infarct may be present (22). In particular, the hypertrophic myocytes of IDCM demonstrate increased DNA (deoxyribonucleic acid) content, increase in the nuclear area with an irregular nuclear membrane, myofibrillar loss, and degenerative changes such as destruction of the sarcomeres (32, 33). Variable levels of chronic inflammation, predominately composed of T-lymphocytes, and increased interstitial collagen are present in hearts with IDCM again, similar to an acute myocardial infarct. According to some studies, interstitial and replacement fibrosis exist in IDCM at frequencies of 15 and 45%, respectively (34, 35). It is important to note that the fibrosis associated with IDCM is characterized by a shift toward oncofetal isoforms of myosin and fibronectin, increased activity of neutrophil-type collagenases, decreased activity of TIMPs accompanied by increased degradation of normal forms of collagen, and increased deposition of abnormal, poorly cross-lined collagen (36, 37).

5. Cardiac Regeneration: Can We Count on Cardiac Myocytes of Cardiac Progenitor Cells?

Mammalian cardiac myocytes withdraw from the cell cycle during the perinatal period and become postmitotic, thus limiting the regenerative capacity of normal human myocardium (38). A few vertebrates such as zebra fish and newts, demonstrate what proponents of myocardial regeneration long for: the ability of cardiac regeneration through initiation of DNA synthesis and cell division (newt), or as in zebrafish, the dedifferentiation of stem cell reserves. Even the human body provides evidence of these desired behaviors, such as in human skeletal muscle, which like the heart does not possess a regenerative capacity. Several hopeful studies suggest that a small amount of myocyte regeneration occurs around infarcted tissue, noting both an increase in mitosis in myocytes of the infarct border zone (39). One theory claims that mitotic myocytes are not terminal differentiated, but instead are suspended in their development with the ability to reenter the cell cycle in response to an infarction. As estimated in one earlier study, approximately 14 myocytes/million are actively undergoing mitosis at any one time. There is an approximately tenfold increase in this number in hearts with end-stage ischemic heart disease (152 myocytes/million) and idiopathic dilated cardiomyopathy (131 myocytes/million). This implies that 81.2×10^3, 882×10^3 and 760×10^3 myocytes are undergoing mitosis in normal, ischemic, and IDCM hearts, respectively, at any one point in time (40). Beltrami et al. used the nuclear antigen Ki-67 to visualize cardiac myocytes undergoing mitosis and found that in the region of the infarct, 4% of the nuclei were undergoing mitosis compared to 1% in the myocyte in regions distant from the myocardium. When measured in terms of a mitotic index (ratio of mitotic to nonmitotic nuclei) myocytes near and distant to the infarct had proliferation indices of 0.08 and 0.03%, respectively (41). Although these findings dispute the assertion that all cardiac myocytes are terminally differentiated, it only supports the claim that cardiac myocytes still retain the capacity for nuclear division where proof of a dividing cardiac stem cell is lacking.

In contrast, von Harsdorf et al. provide three observations to support the currently accepted view that cardiac myocytes do not divide. Their first observation is that cardiac tumors are rare and it has not been proven that these tumors arise from cardiomyocytes. Second, compared to the mitotic indices of mice and amphibians, 10-20% and 15%, respectively, rates of regeneration of human cardiac myocytes are fairly low (as discussed above). Their final observation is that to date there is little evidence that terminally differentiated cardiomyocytes are capable of reaching significant levels of replication.

If terminally differentiated cardiac myocytes cannot replicate, the question regarding the utility of cardiac stem cells remains open. Claims of the evidence for a reserve of cardiac stem cells cite one study in which colonies of cells positive for c-kit, Sca-1, and MDR1 and the Y-chromosome were found in female donor hearts implanted in male recipients. Given that the cells were Y-chromosome positive, there is no dispute that the cells in question were derived from the male recipient and the marker profile was consistent with progenitor cardiac myocytes. These supposed progenitor cells were also observed to lose these markers over time, starting them on path of terminal differentiation and proliferation. It was concluded that terminal differentiation of these progenitor cells was the end result of the process based on the appearance of new myocytes, capillaries, and arterioles in the colonized areas (41). As further evidence, Barile et al. cite a 13-fold increase in cells expressing the same markers in the hypertrophied myocardium of patients with aortic stenosis (38).

Torella et al. suggest the existence of a subset of telomerase positive myocytes in an intermediate state caught between progenitor cells and differentiated myocytes. Within this group is a subgroup of recently derived progenitor, immature, phenotypically smaller cells retaining the ability to enter the cell cycle. These cells are assumed to be responsible for replacing myocytes lost via normal "wear and tear" and there is hope that these "newly born myocytes" could be key to regeneration. These cells have been shown to increase in acute infarcts and with chronic ischemia;

however, one must ask why do we not see more evidence of regeneration. One possible reason is that these "newly born myocytes" cells acquire a senescent phenotype by expressing p16^{INK4a} and subsequently undergo apoptosis and necrosis at a surprisingly higher rate than the surrounding adult myocytes (42).

Despite the aforementioned studies and those similarly searching for answers to cardiac regeneration, it appears that proliferating myocytes are mainly restricted to viable myocardium at the border zone of an infarct (42, 43). In another report, Leri et al. argue that the microenvironment in which cardiac progenitor cells reside may have a significant impact on their ability to differentiate into functioning myocytes. Could it be that regenerating myocytes scattered throughout a background of normally functioning terminally differentiated myocytes have more regenerative capacity than undifferentiated cardiac cells isolated in clusters in the middle of an infarct? It has been shown that myocytes resulting from the differentiation of cardiac progenitor cells are poor versions of adult myocytes as they possess only rare myofibrils and are generally smaller in size compared to the adult phenotype acquired by the same cells when placed among viable myocytes. Further, if neoangiogenesis is taken into account as part of the necessary microenvironment, studies have shown that restoration of blood flow to the area of an infarct is also not sufficient for the regeneration or improved contractility of surrounding myocytes (42, 43).

6. Cell-Based Therapy: Rationale

Without question, adult human hearts exhibit a limited regenerative potential. Therefore, the loss of cardiomyocytes in ischemic heart disease is irreversible where cardiomyocytes that die in response to disease processes or aging are replaced by scar tissue instead of new muscle cells and if replacement fibrosis is extensive enough, the end result is progressive heart failure.

6.1. Stem Cells: A Definition

A broad definition of stem cells relates to a clonogenic population capable of self-renewal and specialized differentiation (44). The two broad categories include adult- and embryonic stem cells (ESCs). Adult stem cells are derived from postnatal somatic tissues and are considered multipotent, meaning they can give rise to multiple differential cells types. In contrast, ESCs, derived from the inner cell mass of blastocyst-stage embryos, are pluripotent with the potential to transform into all differentiated cell types of the postnatal organism. More recently, it was discovered that differentiated somatic cell types can also be reprogrammed into a pluripotent state similar to ESCs via forced expression of stem cell-related genes (iPCS) (45).

6.2. What Cell Types Are Advocated for the Treatment of Myocardial Infarction?

Cardiac differentiation has been reported for cardiac and skeletal myoblasts, and others including adult stem cells isolated from bone marrow-derived, resident cardiac, or peripheral/circulating ECSs. Within these categories, stem cells can be further divided into the number of differentiated cell types they can produce, i.e., the totipotent and the pluripotent cells. The latter may be multipotent or unipotent depending upon the range of differentiated cell lineages (46). We will not address all these issues as they are being addressed in another chapter. Depending on the cell type, each has its unique markers mainly defined by transcription factors and expressed proteins, which potentially aid in the identification when engrafted into target organs. Based on the current literature, however, it is impossible to predict the ideal cell type as most have been reported to mediate at least some functional benefit in preclinical models of cardiac injury.

6.3. Human Pathology and Stem Cell Recognition

To the present, the majority of studies regarding cell-based therapy have focused primarily on animal models with limited clinical data available. The few published studies on the pathology of human stem cells focused mainly on resident cardiac stem cells isolated from ventricular endomyocardial biopsies or surgical material from the atrial appendage.

6.4. Resident Cardiac Stem cells

Previous work has held the notion that the adult heart is a terminally differentiated organ without regenerative capacity where restoration of LV function was overcome by compensation secondary to hypertrophy. The presence of immature cycling cardiomyocytes in human hearts after myocardial injury has raised the possibility that cardiac progenitor cells with regenerative capacity may, in fact, exist (39, 47–49). Cells expressing the stem cell antigen, Sca1, the receptor for stem cells factor, c-kit, the homeodomain transcription factor, islet-1, or undifferentiated cells that grow as self-adherent clusters (cardiospheres) have been identified in the adult heart and have been suggested to be capable of differentiating into cardiomyocytes (Table 1).

In addition to the aforementioned cell types, alternative resident cardiac stem cells with a distinct tissue-specific progenitor cell population have also been identified. These so-termed side population (SP) cells were originally identified by their intrinsic capacity to efflux rhodamine 123 and Hoechst 33342 dye through various ATP-binding cassette (ABC) membrane transporters (50), such as those encoded by multidrug resistance (MDR) genes. Side population cells also harbor telomerase activity, which is only present in replicating cells. The majority of studies regarding the potential SP cells to adopt a more mature cardiac phenotype involve mice where discrepancies in numbers depending on different modes of cardiac injury are found between several

Table 1
Commonly used cardiomyocyte markers and extracardiac tissues/cell types in which they can also be expressed

Marker	Extracardiac site(s) of expression
Nkx2.5	Embryo: pharyngeal endoderm, spleen, stomach, tongue
GATA4	Adult: ovary, testis, lung, liver, and small intestine
	Embryo: proximal and distal gut, testis, ovary, liver, visceral endoderm and parietal endoderm
T-box 5 (Tbx5)	Embryo: eye, forelimb, genital papilla, lung, mandible, trachea
Myocyte enhancer factor	Adult: skeletal muscle, brain, lymphocytes
2C (MEF2c)	Embryo: skeletal muscle, smooth muscle, brain
Connexin43	Ovary, testis, smooth muscle, eye, brain, macrophages, fibroblasts
Sarcomeric MHC	Skeletal muscle
Sarcomeric actin	Skeletal muscle
Sarcomeric actinin	Skeletal muscle
Cardiac troponin I	Fetal skeletal muscle
Cardiac troponin T	Fetal skeletal muscle
Atrial natriuretic peptide	Brain
Smooth muscle α-actin	Smooth muscle, myofibroblasts
Desmin	Smooth muscle, skeletal muscle

Reproduced with permission from ref. (44)

studies (51, 52). Transmembrane ABC-type SP stem cells have been identified in the human heart where ventricular samples from cardiomyopathic heart exhibited significantly increased levels of ABC2 mRNA in dilative and ischemic cardiomyopathy relative to nonfailing hearts (53).

Evidence that resident cardiac-committed cardiac stem cells are well-suited for myocardial regeneration are provided by the laboratory of Marbán et al. in primary cultured endomyocardial biopsy specimens (54). The successful isolation and expansion of cardiac cells raises the possibility that a therapeutically relevant quantity of cells can be generated with appropriate ex vivo protocols (47). These clonogenic cells express stem (CD34, c-kit, sca-1) and endothelial progenitor cell antigens/markers, resembling the adult cardiac lineage. Although proof-of-principle studies exhibit promise for such technology (38), given the minute amount of starting materials and the sorting and expansion of candidate cells (without loss of the cardiac differentiation potential) under Good Manufacturing Practice is likely technically challenging.

6.5. Relevance of c-Kit in Identifying Resident Cardiac Stem Cells

A recent study by the laboratory of Menashé et al., examined 32 endomyocardial biopsies harvested from heart transplant recipients and 18 right appendage biopsies collected during coronary artery bypass surgery (55). The tissues were immunostained for detection markers of stemness (namely, c-kit, MDR-1, Isl-1), hematopoietic origin (CD45), mast cells (tryptase), endothelial cells (CD105), and cardiac lineage (Nkx2.5). A median number of c-kit positive cells/mm^2 (2.7 [1.8–4 c-kit$^+$]) were detected for endomyocardial biopsies, while reduced numbers were found in the right appendage (2 [0.5–1.8 c-kit$^+$, $p=0.01$]). Despite the findings of c-kit immunoexpression, other markers of stemness or cardiac differentiation pathways were not detected. Moreover, the majority of c-kit positive cells were found in a perivascular location consistent with a mast cell phenotype positive for both CD45 and tryptase. Similar findings of a lack of cardiac expression markers in c-kit-positive cells were report by Yamabi and colleagues in cells isolated from human fetal hearts (56).

The lack of expression of stemness or cardiogenic markers as demonstrated in human endomyocardial or appendage biopsies raises important questions regarding the validity of markers selected to assess myocardial regeneration. For example, consideration must be given to the possibility of phenotypic changes induced by repeat passages of cells in culture such as in the experiments by the laboratory of Marbán (38). Moreover, markers of cardiac differentiation such as Sca-1 are expressed only after exposure to agents such as 5-azacytidine (57), therefore, its expression would be negative in tissues or primary cells. The type of cell induction media used in culture is another consideration since expressed phenotypes are likely affected. Furthermore, confirmatory evidence of cardiac regeneration is complicated by fusion of engrafted cells with the host tissue (58–60). The benefits of cell grafting may not be exclusive or even related to the formation of new myocardium, but may arise from secondary effects associated with postinfarct remodeling or increased angiogenesis.

Identification and tracking of stem cells can be achieved by immune phenotyping using immunohistochemistry and/or flow cytometry, or emerging newer genomic or proteomic approaches. A list of divergent adult progenitor cell types with cardiac repair potential and their respective markers are summarized in Table 2. Common strategies to determine the cardiomyogenic traits of progenitor cells include fate mapping using chemical labels (BrdU, fluorescent dyes, iron particles), gender-specific DNA sequences, (Y-chromosome), reporter genes (GFP), and independent genetic reporters (44). Molecular studies using species-specific RT-PCR for cardiac transcripts can be used as simple high-throughput assays. The functional component of the graft (electromechanical integration) can be assessed directly by intravital calcium imaging of GFP-tagged grafts while overall assessment of cardiac function

Table 2
The four types of resident cardiac stem and progenitor cells identified so far in humans and their salient characteristics

Characteristic	Type of cell			
	c-Kit	Sca-1	MDR-1	Isl-1
Cardiac differentiation	Yes	Yes	Yes	Yes
Self-renewal	Yes	Yes	Yes	Yes
Clonogenic	Yes	Yes	Yes	Not known
Multipotent	Yes	Yes	Yes	Not known
Cardiosphere formation[a]	Yes	Yes	Yes	Not known
Present in adult/fetus	Both	Both	Both	Fetus to adult

[a]Pseudo-embryoid bodies (a marker of multipotency) when cells are grown in suspension
Reproduced with permission from ref. (42)

can be achieved by routine echocardiography. Despite the diversity of markers and techniques, the existence of these methodologies in no way provides conclusive evidence of specific cell transformation into the cardiac lineage where data can only be interpreted as supportive to the fact.

6.6. Most Cells with a Destined Cardiac Lineage Improve Myocardial Performance

One principal factor undermining the clinical success of cell therapy is the necessity that a substantial fraction of the injured myocardium be replaced in order to truly overcome the deficit created by injured or ischemic myocardium. As a practical matter, Olivetti et al. showed that the adult myocardium consists of approximately 20 million cardiomyocytes per gram of tissue (61). Thus, the average left ventricle weighing approximately 200 g would contain approximately four billion myocytes. Damage to at least 25% of the left ventricle causes heart failure with a risk for cardiogenic shock when this injury exceeds 40%. Therefore, one can surmise that primary remuscularization might require billions of cells to replace the deficit observed in the failing heart.

The restoration of a significant amount of myocardium as a beating syncytium certainly becomes a fundamental issue when ectopically placed committed myogenic cell sources such as skeletal myoblasts, fetal/neonatal cardiomyocytes or even more divergent cells types such as endothelial progenitor cells (EPCs), mesenchymal stem cells (MSCs), resident cardiac progenitor cells, and ESCs are implemented for cell-based cardiac repair. Although stem cells are reported to provide a physiologic benefit, there remains a lack of mechanistic understanding as to the ideal cell type, cell number, means and timing of delivery in the postinfarction myocardium. Moreover, any treatment benefit to date is likely derived from other sources than engrafted cells beating in synchrony with the host myocardium (62). The limited regenerative capacity of the human heart, unlike zebra fish or the newt, simply arises from its inability to form daughter cardiomyocytes, with even less capability for repair in infarcted or cardiomyopathic processes.

Indeed, the objectives of cell-based interventions can be directed toward a paracrine or structural approach where the latter would require a substantially greater number of cells. Reminiscent of diabetes or Parkinson disease, the paracrine objective supposes that the engrafted cells primarily supply a missing mediator, such as insulin or dopamine, respectively. In this setting, the adopted phenotype of the replacement cells does not have to precisely match the diseased host unlike the structural objective, which implies that the grafted cells support true regeneration of a dead tissue. Thus, in the context of heart failure, improvement of left ventricular function would require that a large area of dysfunctional myocardium be repopulated by new cells capable of contractile activity along with the ability to electromechanically couple with host cardiomyocytes.

6.7. Recognition of the Newly-Differentiated Cardiomyocyte

Any newly generated myocyte would need to show positivity for several cardiac myocyte-specific markers including cardiac troponin I, sarcomeric actins and myosins, myosin light chain 2V, GATA-4, and ANF (see Table 3). Functioning myocytes will show positivity for proteins involved in electrophysiologic coupling of adjoining cells, such as N-cadherin and connexin43. Important negative stains include vimentin and procollagen, which will stain fibroblasts only (63).

6.8. Potential Microscopic Artifacts in the Recognition of Stem Cell-Derived Myocytes

Some common artifacts of circulating progenitors include the accumulation of resident leukocytes and the high intrinsic autofluorescence of infarcted myocardial tissue. Autofluorescence is a recognized problem of normal striated muscle and can be mistaken for transgene expression of enhanced green fluorescence protein (EGFP) or a fluorescent immunostain (64). Following injury, autofluorescence increases as a result of accumulated lipofuscin, blood-derived pigments and other intrinsic fluors including flavins and reduced nicotinamide adenine dinucleotide (NADH). In the case of EGFP, confusion with autofluorescence can be minimized by immunolabeling with specific anti-EGFP antibody, rather than relying on intrinsic fluorescence alone. The advantage of an immunohistochemical approach is the apparent amplification of the signal thereby reducing the signal-to-background ratio by ~300-fold (63). It is essential to stain samples in the presence of a well-established positive and negative control along with determining the background fluorescence generated

Table 3
Adult stem cells and some commonly expressed markers

Adult stem cells	Markers
Bone marrow-derived stem cells	
Hematopoietic stem cells (HSCs)	CD34+/– (73, 74), CD45 (75)
Mesenchymal stem cells (MSCs)	CD29, CD44, CD71, CD90, CD106, CD120a, CD124 (44), CD73, CD105, CD166 (76)
Multipotent adult progenitor cells (MAPCs)	GFP, FLK-1$^+$, VEGFR-2, Sca-1, CD13 (77); Oct 3/4 (POU5F1), FLK-1$^-$ (78)
Bone-marrow-derived stem cells (BMSCs)	CD13, CD29, CD44, CD90, CD105, CD166 (79)
Very small embryonic-like stem cells (VSESCs)	Lin$^-$CXCR4$^+$, CD45$^-$, coexpress CD133 and CD34 antigen (80)
Lin$^-$CXCR4$^+$ CD45$^-$, coexpress CD133 and CD34 antigen (80)	
Adipose-derived stem cells (ADSCs)	CD45$^-$, CD31$^-$, CD34$^+$ and CD13$^+$, CD29$^+$, CD 44$^+$, CD105$^+$, and CD166$^+$ (79)
Endothelial progenitor cells (EPCs)	Fkl-1, CD34, CD133 (44)
Umbilical cord-derived stem cells (UCBSCs)	CD13, CD29, CD44, CD90, CD105, CD166 (81)

by the secondary antibody. The intense background autofluorescence in formaldehyde-fixed myocardial sections may be further controlled by exposure to 0.1% Sudan Black B after fluorescence staining (65). Finally, the engulfment of necrotic cells by leukocytes may also confuse the recognition of stem cell-derived myocytes since these cells may show remnants of myocyte nuclei and/or cardiac myofibrils (64). Inflammatory cell markers such as CD45 and CD68 or further characterization by transmission electron microscopy may help rule out the presence of leukocytes.

6.9. Complications of Myocyte Fusion

Myocardial regeneration could result from fusion of the transplanted stem cell with existing cardiac cells resulting in the formation of hybrid cells. The formation of a hybrid cell may contribute to the development or maintenance of these key cell types. Cell fusion should generate binucleated myocytes with one tetraploid and one diploid nucleus, or myocytes with three diploid nuclei. Evidence for fusion between cardiac and stem cells has been reported using selective fluorophore low molecular weight tracers such as calcein-AM (66), nanocrystal labeling (67), combined

EGFP and transgender cardiac transplantation (63), and Cre/ Lox donor/recipient pairs (57, 58, 68). In the latter method, donor stem cells labeled with a reporter gene are injected into hearts of mice in which Cre recombinase (a site-specific DNA recombinase) is expressed only in cardiomyocytes. If fusion occurs between stem cells and cardiomyocytes, the Cre recombinase will excise a floxed stop sequence allowing the expression of a reporter gene. The cre-lox genetic system technique, however, is not perfect and may generate false positive results through metabolic cooperation (63) where a cell acquires the cre-recombinase from a neighboring cell and undergoes excision of the flox-flanked DNA segment in the absence of cell fusion presumably by an exchange of the enzyme through intercellular junctions (63).

Myocyte volume size is another potential determinant of cell fusion with expected new myocytes of equal or greater volume (63). Although the incidence of cell fusion in the postinfarcted myocardium is not well defined, critical evaluation of the current reports indicate that cell fusion is not a dominant (or perhaps nonexistent) mechanism and may be considered essentially an in vitro phenomenon with few implications in vivo (69).

Demonstrating an unmistakable cardiac phenotype can be challenging where more definitive morphology can be confirmed by immunostaining using one or more cardiac-specific markers and/or by RT-PCR. In reality, there is no single precise cardiac marker, and so the approach requires a thoughtful selection of multiple markers with appropriate positive and negative controls.

6.10. Potential Complications of Cell-Based Therapy

Like most controversial therapies, valid concerns are raised about potential adverse effects, such as tumorigenicity or unregulated differentiation after stem cell implantation. Along these lines, documented teratoma formation after implantation of undifferentiated ESC in knee joints (70) and hearts of immunotolerant hosts have been reported (64). The incidence of intramyocardial teratomas, however, is rare with only six reported cases in humans (71) and one additional case in the Armed Forces Institute of Pathology files where most occur in newborns during the first 6 years of life. The majority of patients with cardiac teratomas present with congestive heart failure while sudden death precipitated from acute arrhythmias caused by the tumor's interventricular location may be a first symptom (71).

Other reported complications of experimental cell-based therapy involve ventricular calcification after local transplantation of unselected bone marrow cells in AMI in rats (72), although the clinical consequences of severe ventricular calcification in the post-MI patient is unknown. Further complications of cell transplantation in the heart include arrhythmias, coronary plaque angiogenesis with plaque rupture, possible hemorrhagic pericarditis, and infarct extension.

7. Summary
Perspectives

Acute or healed myocardial infarction remains the primary cause of high morbidity and mortality in patients presenting with acute coronary syndromes or congestive heart failure. This reality creates a search for newer therapies directed at treating myocardial infarction beyond pharmacologic agents alone. Cell-based transplantation, at least in animals, provides a proof-of-principle approach of treating AMI by repopulating lost myocytes and/or induction of paracrine effects. The effectiveness of cell therapy, however, requires careful monitoring of cell plasticity in addition to the ability of transplanted cells to integrate into a functional myocardium and cognizance of potential artifacts that may mislead results. Critical issues as to the utility of differentiated or multipotent cell types are yet unresolved and universal identification of markers that recognize optimal stem cell types are yet unavailable. Moreover, it must be emphasized that establishment of viable cell-derived myocytes without functional improvement is of no clinical significance toward the optimal goal of providing a long-term therapeutic benefit with minimal risk. This rapidly paced field is in its infancy and serious inroads into providing clinically effective cell-based therapy will likely require a more comprehensive understanding of regenerative biology.

References

1. Thom, T., Haase, N., Rosamond, W., Howard, V. J., Rumsfeld, J., Manolio, T., Zheng, Z. J., Flegal, K., O'Donnell, C., Kittner, S., Lloyd-Jones, D., Goff, D. C., Jr., Hong, Y., Adams, R., Friday, G., Furie, K., Gorelick, P., Kissela, B., Marler, J., Meigs, J., Roger, V., Sidney, S., Sorlie, P., Steinberger, J., Wasserthiel-Smoller, S., Wilson, M., and Wolf, P. (2006) Heart disease and stroke statistics – 2006 update: a report from the American Heart Association Statistics Committee and Stroke Statistics Subcommittee, Circulation 113, e85–e151.

2. Hunt, S. A., Baker, D. W., Chin, M. H., Cinquegrani, M. P., Feldman, A. M., Francis, G. S., Ganiats, T. G., Goldstein, S., Gregoratos, G., Jessup, M. L., Noble, R. J., Packer, M., Silver, M. A., Stevenson, L. W., Gibbons, R. J., Antman, E. M., Alpert, J. S., Faxon, D. P., Fuster, V., Jacobs, A. K., Hiratzka, L. F., Russell, R. O., and Smith, S. C. Jr. (2001). ACC/AHA guidelines for the evaluation and management of chronic heart failure in the adult: executive summary. A report of the American College of Cardiology/American Heart Association Task Force on Practice Guidelines (Committee to revise the 1995 Guidelines for the Evaluation and Management of Heart Failure), J Am Coll Cardiol 38, 2101–2113.

3. Mariani, J., Ou, R., Bailey, M., Rowland, M., Nagley, P., Rosenfeldt, F., and Pepe, S. (2000) Tolerance to ischemia and hypoxia is reduced in aged human myocardium, J Thorac Cardiovasc Surg 120, 660–667.

4. Juhaszova, M., Rabuel, C., Zorov, D. B., Lakatta, E. G., and Sollott, S. J. (2005) Protection in the aged heart: preventing the heart-break of old age?, Cardiovasc Res 66, 233–244.

5. Mosterd, A., and Hoes, A. W. (2007) Clinical epidemiology of heart failure, Heart 93, 1137–1146.

6. Sedmera, D. (2005) Form follows function: developmental and physiological view on ventricular myocardial architecture, Eur J Cardiothorac Surg 28, 526–528.

7. Wenink, A. C., and Gittenberger-de Groot, A. C. (1982) Left and right ventricular trabecular patterns. Consequence of ventricular

septation and valve development, *Br Heart J 48*, 462–468.

8. Sedmera, D., Pexieder, T., Vuillemin, M., Thompson, R. P., and Anderson, R. H. (2000) Developmental patterning of the myocardium, *Anat Rec 258*, 319–337.

9. Chien, K. R., Domian, I. J., and Parker, K. K. (2008) Cardiogenesis and the complex biology of regenerative cardiovascular medicine, *Science 322*, 1494–1497.

10. Kelly, R. G., Brown, N. A., and Buckingham, M. E. (2001) The arterial pole of the mouse heart forms from Fgf10-expressing cells in pharyngeal mesoderm, *Dev Cell 1*, 435–440.

11. Buckingham, M., Meilhac, S., and Zaffran, S. (2005) Building the mammalian heart from two sources of myocardial cells, *Nat Rev Genet 6*, 826–835.

12. Laugwitz, K. L., Moretti, A., Lam, J., Gruber, P., Chen, Y., Woodard, S., Lin, L. Z., Cai, C. L., Lu, M. M., Reth, M., Platoshyn, O., Yuan, J. X., Evans, S., and Chien, K. R. (2005) Postnatal isl1+ cardioblasts enter fully differentiated cardiomyocyte lineages, *Nature 433*, 647–653.

13. Moretti, A., Caron, L., Nakano, A., Lam, J. T., Bernshausen, A., Chen, Y., Qyang, Y., Bu, L., Sasaki, M., Martin-Puig, S., Sun, Y., Evans, S. M., Laugwitz, K. L., and Chien, K. R. (2006) Multipotent embryonic isl1+ progenitor cells lead to cardiac, smooth muscle, and endothelial cell diversification, *Cell 127*, 1151–1165.

14. Cai, C. L., Liang, X., Shi, Y., Chu, P. H., Pfaff, S. L., Chen, J., and Evans, S. (2003) Isl1 identifies a cardiac progenitor population that proliferates prior to differentiation and contributes a majority of cells to the heart, *Dev Cell 5*, 877–889.

15. Rychter, Z., and Ostadal, B. (1971) Fate of "sinusoidal" intertrabecular spaces of the cardiac wall after development of the coronary vascular bed in chick embryo, *Folia Morphol (Praha) 19*, 31–44.

16. Varnava, A. M. (2001) Isolated left ventricular non-compaction: a distinct cardiomyopathy?, *Heart 86*, 599–600.

17. de Jong, F., Opthof, T., Wilde, A. A., Janse, M. J., Charles, R., Lamers, W. H., and Moorman, A. F. (1992) Persisting zones of slow impulse conduction in developing chicken hearts, *Circ Res 71*, 240–250.

18. Reckova, M., Rosengarten, C., deAlmeida, A., Stanley, C. P., Wessels, A., Gourdie, R. G., Thompson, R. P., and Sedmera, D. (2003) Hemodynamics is a key epigenetic factor in development of the cardiac conduction system, *Circ Res 93*, 77–85.

19. Kumar, V., Abbas, A. K., and Fausto, N., (Ed.) (2005) *Robbins and Cotran Pathological Basis of Disease*, 7th ed., Elsevier Saunders, Philadelphia.

20. Stoker, M. E., Gerdes, A. M., and May, J. F. (1982) Regional differences in capillary density and myocyte size in the normal human heart, *Anat Rec 202*, 187–191.

21. Miura, T., Shizukuda, Y., Ogawa, S., Ishimoto, R., and Iimura, O. (1991) Effects of early and later reperfusion on healing speed of experimental myocardial infarct, *Can J Cardiol 7*, 146–154.

22. Burke, A. P., and Virmani, R. (2007) Pathophysiology of acute myocardial infarction, *Med Clin North Am 91*, 553–572; ix.

23. Virag, J. I., and Murry, C. E. (2003) Myofibroblast and endothelial cell proliferation during murine myocardial infarct repair, *Am J Pathol 163*, 2433–2440.

24. Sun, Y., Kiani, M. F., Postlethwaite, A. E., and Weber, K. T. (2002) Infarct scar as living tissue, *Basic Res Cardiol 97*, 343–347.

25. Beltrami, C. A., Finato, N., Rocco, M., Feruglio, G. A., Puricelli, C., Cigola, E., Quaini, F., Sonnenblick, E. H., Olivetti, G., and Anversa, P. (1994) Structural basis of end-stage failure in ischemic cardiomyopathy in humans, *Circulation 89*, 151–163.

26. Virmani, R., Kolodgie, F. D., and Ladich, E. (2006) Mechanistic insights into cardiac stem cell therapy, in *A Guide to Cardiac Cell Therapy* (Perlin, E., and Willerson, J. T., Eds.) 2 ed., Informa HealthCare, Milton Park, Abingdon, UK.

27. Jugdutt, B. I. (2003) Ventricular remodeling after infarction and the extracellular collagen matrix: when is enough enough?, *Circulation 108*, 1395–1403.

28. Alberts, B., Bray, D., Lewis, J., Raff, M., Roberts, K., and Watson, J. D. (1994) *Molecular Biology of the Cell*, 3rd ed., Garland Publishing, New York.

29. Philips, C., and Wenstrup, R. J., (Ed.) (1992) *Biosynthetic and Genetic Disorders of Collagen*, Saunders, Philadelphia.

30. van den Borne, S. W., Cleutjens, J. P., Hanemaaijer, R., Creemers, E. E., Smits, J. F., Daemen, M. J., and Blankesteijn, W. M. (2009) Increased matrix metalloproteinase-8 and -9 activity in patients with infarct rupture after myocardial infarction, *Cardiovasc Pathol 18*, 37–43.

31. Ertl, G., and Frantz, S. (2005) Healing after myocardial infarction, *Cardiovasc Res 66*, 22–32.

32. Unverferth, D. V., Baker, P. B., Swift, S. E., Chaffee, R., Fetters, J. K., Uretsky, B. F., Thompson, M. E., and Leier, C. V. (1986) Extent of myocardial fibrosis and cellular hypertrophy in dilated cardiomyopathy, *Am J Cardiol 57*, 816–820.

33. Schwarz, F., Mall, G., Zebe, H., Blickle, J., Derks, H., Manthey, J., and Kubler, W. (1983) Quantitative morphologic findings of the myocardium in idiopathic dilated cardiomyopathy, *Am J Cardiol 51*, 501–506.

34. Roberts, W. C., Siegel, R. J., and McManus, B. M. (1987) Idiopathic dilated cardiomyopathy: analysis of 152 necropsy patients, *Am J Cardiol 60*, 1340–1355.

35. Rose, A. G., and Beck, W. (1985) Dilated (congestive) cardiomyopathy: a syndrome of severe cardiac dysfunction with remarkably few morphological features of myocardial damage, *Histopathology 9*, 367–379.

36. Gabler, U., Berndt, A., Kosmehl, H., Mandel, U., Zardi, L., Muller, S., Stelzner, A., and Katenkamp, D. (1996) Matrix remodelling in dilated cardiomyopathy entails the occurrence of oncofetal fibronectin molecular variants, *Heart 75*, 358–362.

37. Gunja-Smith, Z., Morales, A. R., Romanelli, R., and Woessner, J. F., Jr. (1996) Remodeling of human myocardial collagen in idiopathic dilated cardiomyopathy. Role of metalloproteinases and pyridinoline cross-links, *Am J Pathol 148*, 1639–1648.

38. Barile, L., Messina, E., Giacomello, A., and Marbán, E. (2007) Endogenous cardiac stem cells, *Prog Cardiovasc Dis 50*, 31–48.

39. Beltrami, A. P., Urbanek, K., Kajstura, J., Yan, S. M., Finato, N., Bussani, R., Nadal-Ginard, B., Silvestri, F., Leri, A., Beltrami, C. A., and Anversa, P. (2001) Evidence that human cardiac myocytes divide after myocardial infarction, *N Engl J Med 344*, 1750–1757.

40. Kajstura, J., Leri, A., Finato, N., Di Loreto, C., Beltrami, C. A., and Anversa, P. (1998) Myocyte proliferation in end-stage cardiac failure in humans, *Proc Natl Acad Sci U S A 95*, 8801–8805.

41. Quaini, F., Urbanek, K., Beltrami, A. P., Finato, N., Beltrami, C. A., Nadal-Ginard, B., Kajstura, J., Leri, A., and Anversa, P. (2002) Chimerism of the transplanted heart, *N Engl J Med 346*, 5–15.

42. Torella, D., Ellison, G. M., Mendez-Ferrer, S., Ibanez, B., and Nadal-Ginard, B. (2006) Resident human cardiac stem cells: role in cardiac cellular homeostasis and potential for myocardial regeneration, *Nat Clin Pract Cardiovasc Med 3 Suppl 1*, S8–S13.

43. Urbanek, K., Quaini, F., Tasca, G., Torella, D., Castaldo, C., Nadal-Ginard, B., Leri, A., Kajstura, J., Quaini, E., and Anversa, P. (2003) Intense myocyte formation from cardiac stem cells in human cardiac hypertrophy, *Proc Natl Acad Sci U S A 100*, 10440–10445.

44. Reinecke, H., Minami, E., Zhu, W. Z., and Laflamme, M. A. (2008) Cardiogenic differentiation and transdifferentiation of progenitor cells, *Circ Res 103*, 1058–1071.

45. Takahashi, K., and Yamanaka, S. (2006) Induction of pluripotent stem cells from mouse embryonic and adult fibroblast cultures by defined factors, *Cell 126*, 663–676.

46. Mathur, A., and Martin, J. F. (2004) Stem cells and repair of the heart, *Lancet 364*, 183–192.

47. Messina, E., De Angelis, L., Frati, G., Morrone, S., Chimenti, S., Fiordaliso, F., Salio, M., Battaglia, M., Latronico, M. V., Coletta, M., Vivarelli, E., Frati, L., Cossu, G., and Giacomello, A. (2004) Isolation and expansion of adult cardiac stem cells from human and murine heart, *Circ Res 95*, 911–921.

48. Srivastava, D., and Olson, E. N. (2000) A genetic blueprint for cardiac development, *Nature 407*, 221–226.

49. Hirschmann-Jax, C., Foster, A. E., Wulf, G. G., Nuchtern, J. G., Jax, T. W., Gobel, U., Goodell, M. A., and Brenner, M. K. (2004) A distinct "side population" of cells with high drug efflux capacity in human tumor cells, *Proc Natl Acad Sci U S A 101*, 14228–14233.

50. Hierlihy, A. M., Seale, P., Lobe, C. G., Rudnicki, M. A., and Megeney, L. A. (2002) The post-natal heart contains a myocardial stem cell population, *FEBS Lett 530*, 239–243.

51. Martin, C. M., Ferdous, A., Gallardo, T., Humphries, C., Sadek, H., Caprioli, A., Garcia, J. A., Szweda, L. I., Garry, M. G., and Garry, D. J. (2008) Hypoxia-inducible factor-2alpha transactivates Abcg2 and promotes cytoprotection in cardiac side population cells, *Circ Res 102*, 1075–1081.

52. Mouquet, F., Pfister, O., Jain, M., Oikonomopoulos, A., Ngoy, S., Summer, R., Fine, A., and Liao, R. (2005) Restoration of cardiac progenitor cells after myocardial infarction by self-proliferation and selective homing of bone marrow-derived stem cells, *Circ Res 97*, 1090–1092.

53. Meissner, K., Heydrich, B., Jedlitschky, G., Meyer Zu Schwabedissen, H., Mosyagin, I., Dazert, P., Eckel, L., Vogelgesang, S., Warzok,

R. W., Bohm, M., Lehmann, C., Wendt, M., Cascorbi, I., and Kroemer, H. K. (2006) The ATP-binding cassette transporter ABCG2 (BCRP), a marker for side population stem cells, is expressed in human heart, *J Histochem Cytochem 54*, 215–221.

54. Smith, R. R., Barile, L., Cho, H. C., Leppo, M. K., Hare, J. M., Messina, E., Giacomello, A., Abraham, M. R., and Marbán, E. (2007) Regenerative potential of cardiosphere-derived cells expanded from percutaneous endomyocardial biopsy specimens, *Circulation 115*, 896–908.

55. Pouly, J., Bruneval, P., Mandet, C., Proksch, S., Peyrard, S., Amrein, C., Bousseaux, V., Guillemain, R., Deloche, A., Fabiani, J. N., and Menasche, P. (2008) Cardiac stem cells in the real world, *J Thorac Cardiovasc Surg 135*, 673–678.

56. Yamabi, H., Lu, H., Dai, X., Lu, Y., Hannigan, G., and Coles, J. G. (2006) Overexpression of integrin-linked kinase induces cardiac stem cell expansion, *J Thorac Cardiovasc Surg 132*, 1272–1279.

57. Oh, H., Bradfute, S. B., Gallardo, T. D., Nakamura, T., Gaussin, V., Mishina, Y., Pocius, J., Michael, L. H., Behringer, R. R., Garry, D. J., Entman, M. L., and Schneider, M. D. (2003) Cardiac progenitor cells from adult myocardium: homing, differentiation, and fusion after infarction, *Proc Natl Acad Sci U S A 100*, 12313–12318.

58. Alvarez-Dolado, M., Pardal, R., Garcia-Verdugo, J. M., Fike, J. R., Lee, H. O., Pfeffer, K., Lois, C., Morrison, S. J., and Alvarez-Buylla, A. (2003) Fusion of bone-marrow-derived cells with Purkinje neurons, cardiomyocytes and hepatocytes, *Nature 425*, 968–973.

59. Terada, N., Hamazaki, T., Oka, M., Hoki, M., Mastalerz, D. M., Nakano, Y., Meyer, E. M., Morel, L., Petersen, B. E., and Scott, E. W. (2002) Bone marrow cells adopt the phenotype of other cells by spontaneous cell fusion, *Nature 416*, 542–545.

60. Ying, Q. L., Nichols, J., Evans, E. P., and Smith, A. G. (2002) Changing potency by spontaneous fusion, *Nature 416*, 545–548.

61. Olivetti, G., Capasso, J. M., Sonnenblick, E. H., and Anversa, P. (1990) Side-to-side slippage of myocytes participates in ventricular wall remodeling acutely after myocardial infarction in rats, *Circ Res 67*, 23–34.

62. Murry, C. E., Reinecke, H., and Pabon, L. M. (2006) Regeneration gaps: observations on stem cells and cardiac repair, *J Am Coll Cardiol 47*, 1777–1785.

63. Kajstura, J., Rota, M., Whang, B., Cascapera, S., Hosoda, T., Bearzi, C., Nurzynska, D., Kasahara, H., Zias, E., Bonafe, M., Nadal-Ginard, B., Torella, D., Nascimbene, A., Quaini, F., Urbanek, K., Leri, A., and Anversa, P. (2005) Bone marrow cells differentiate in cardiac cell lineages after infarction independently of cell fusion, *Circ Res 96*, 127–137.

64. Laflamme, M. A., and Murry, C. E. (2005) Regenerating the heart, *Nat Biotechnol 23*, 845–856.

65. Baschong, W., Suetterlin, R., and Laeng, R. H. (2001) Control of autofluorescence of archival formaldehyde-fixed, paraffin-embedded tissue in confocal laser scanning microscopy (CLSM), *J Histochem Cytochem 49*, 1565–1572.

66. Driesen, R. B., Dispersyn, G. D., Verheyen, F. K., van den Eijnde, S. M., Hofstra, L., Thone, F., Dijkstra, P., Debie, W., Borgers, M., and Ramaekers, F. C. (2005) Partial cell fusion: a newly recognized type of communication between dedifferentiating cardiomyocytes and fibroblasts, *Cardiovasc Res 68*, 37–46.

67. Murasawa, S., Kawamoto, A., Horii, M., Nakamori, S., and Asahara, T. (2005) Niche-dependent translineage commitment of endothelial progenitor cells, not cell fusion in general, into myocardial lineage cells, *Arterioscler Thromb Vasc Biol 25*, 1388–1394.

68. Reinecke, H., Minami, E., Poppa, V., and Murry, C. E. (2004) Evidence for fusion between cardiac and skeletal muscle cells, *Circ Res 94*, e56–e60.

69. Harris, R. G., Herzog, E. L., Bruscia, E. M., Grove, J. E., Van Arnam, J. S., and Krause, D. S. (2004) Lack of a fusion requirement for development of bone marrow-derived epithelia, *Science 305*, 90–93.

70. Wakitani, S., Takaoka, K., Hattori, T., Miyazawa, N., Iwanaga, T., Takeda, S., Watanabe, T. K., and Tanigami, A. (2003) Embryonic stem cells injected into the mouse knee joint form teratomas and subsequently destroy the joint, *Rheumatology (Oxford) 42*, 162–165.

71. Swalwell, C. I. (1993) Benign intracardiac teratoma. A case of sudden death, *Arch Pathol Lab Med 117*, 739–742.

72. Yoon, Y. S., Park, J. S., Tkebuchava, T., Luedeman, C., and Losordo, D. W. (2004) Unexpected severe calcification after transplantation of bone marrow cells in acute myocardial infarction, *Circulation 109*, 3154–3157.

73. Pei, X. (1999) Who is hematopoietic stem cell: CD34+ or CD34–? *Int J Hematol 70*, 213–215.

74. Zhao, T. C., Tseng, A., Yano, N., et al. (2008) Targeting human CD34+ hematopoietic stem cells with anti-CD45 × anti-myosin light-chain bispecific antibody preserves cardiac function in myocardial infarction. *J Appl Physiol 104*, 1793–1800.

75. Narasipura, S. D., Wojciechowski, J. C., Duffy, B. M., L Liesveld, J., King, M. R. (2008) Purification of CD45+ hematopoietic cells directly from human bone marrow using a flow-based P-selectin-coated microtube. *Am J Hematol 83*, 627–629.

76. Ryan, J. M., Barry, F. P., Murphy, J. M., and Mahon, B. P. (2005) Mesenchymal stem cells avoid allogeneic rejection. *J Inflamm (Lond) 2*, 8.

77. Reyes, M., Li, S., Foraker, J., Kimura, E., Chamberlain, J. S. (2005) Donor origin of multipotent adult progenitor cells in radiation chimeras, *Blood 106*, 3646–3649.

78. Ross, J. J., Hong, Z., Willenbring, B., et al. (2006) Cytokine-induced differentiation of multipotent adult progenitor cells into functional smooth muscle cells. *J Clin Invest 116*, 3139–3149.

79. Zhu, Y., Liu, T., Song, K., Fan, X., Ma, X., and Cui, Z. (2008) Adipose-derived stem cell: a better stem cell than BMSC. *Cell Biochem Funct 26*, 664–675.

80. Zuba-Surma, E. K., Kucia, M., Ratajczak, J., Ratajczak, M. Z. (2009) "Small stem cells" in adult tissues: very small embryonic-like stem cells stand up! *Cytometry A 75*, 4–13.

81. Wu, K. H., Yang, S. G., Zhou, B., et al. (2007) Human umbilical cord derived stem cells for the injured heart. *Med Hypotheses 68*, 94–97.

Chapter 10

Assessment of Myocardial Angiogenesis and Vascularity in Small Animal Models

Matthew L. Springer

Abstract

Therapies that aim to prevent myocardial tissue from dying or to regenerate new myocardium all rely on the preservation or growth of a functional vasculature. The amount of blood that supplies the myocardium is dependent on the number and nature of the microvessels, as well as the ability of the arteries to supply blood and the veins to remove it. All of these factors can be assessed when success of an experimental therapy is being evaluated. Different kinds of information can be obtained from these different parameters, and it is important to understand what each one involves and how it can be misinterpreted. This chapter describes the various approaches to the assessment of vascularity in the heart with a focus on small animal models, dealing both with those approaches that are purely histological endpoint studies and those that are functional measurements in living animals.

Key words: Vascularity, Angiogenesis, Neovascularization, Capillaries, Length density, Area density, Perfusion, Myocardial infarction, Myocardial regeneration

1. Introduction

In the development of therapies for ischemic myocardial disease, one might say that all roads lead to the vasculature. If tissue is ischemic due to chronic poor circulation or an acute event like a myocardial infarction (MI), the augmentation of the preexisting vasculature can theoretically rescue tissue that may otherwise succumb to ischemic death. At the other extreme, the use of various tissue engineering approaches, including cell therapy to regenerate tissue that has already died, will require that the new tissue be sufficiently vascularized to supply blood. Regardless of the circumstance and the therapeutic approach, assessment of

Randall J. Lee (ed.), *Stem Cells for Myocardial Regeneration: Methods and Protocols*, Methods in Molecular Biology, vol. 660, DOI 10.1007/978-1-60761-705-1_10, © Springer Science+Business Media, LLC 2010

vascularity is one of the most common, most bedeviling, and most easily misinterpreted techniques encountered in the literature.

This chapter will compare several common approaches toward histological and functional assessment of vascularity. Subheadings 2 and 3 together present a basic protocol for histological staining of capillaries and arterioles in tissue sections, but the emphasis of this chapter is in Subheading 4, which deals with several important issues and approaches. Some readers may be surprised to realize that merely "counting vessels" is not straightforward; it means different things to different people and is, in some cases, not helpful.

2. Materials

1. Phosphate-buffered saline (PBS) – A convenient, reasonably economical alternative to making it from scratch is to use premade tablets that are dissolved in water – Sigma #P4417-100TAB.

2. 1.5% formaldehyde in PBS – Dilute from "16% paraformaldehyde" ampules – Electron Microscopy Sciences #15710.
   This is frequently but inaccurately labeled 16% paraformaldehyde (paraformaldehyde breaks down to formaldehyde in aqueous solution). Using ampules is a safer alternative to dissolving paraformaldehyde powder. Solution is best fresh but does not noticeably decline in quality for light level microscopy during 2 weeks at 4°C.

3. PAP pen, Super HT – Research Products International Corp. (no product number).

4. Staining buffer – make in PBS:
   - 2% normal goat serum.
   - 0.3% Triton X-100.
   - 0.02% sodium azide.

   All reagents available from Sigma, can be stored for several months at 4°C. Triton X-100 is extremely viscous; measure by weight, or dilute from a 10% stock solution in water.

5. Mouse-on-mouse block (if staining mouse tissue) – dilute in staining buffer:
   - AffiniPure Fab fragment from goat antimouse IgG – Jackson ImmunoResearch #115-007-003.
   - ChromPure goat IgG, Fc fragment – Jackson ImmunoResearch #005-000-008.

   This is adapted from reference (1). In our experience, this mouse-on-mouse block works better than commercially available reagents advertised as doing the same thing.

6. Biotin Blocking System – Dako #X0590.

7. BS-1 isolectin B_4, biotin conjugate, from *Griffonia simplicifolia* (*Bandeiraea simplicifolia*) – Sigma #L2140.

8. Mouse anti-α-smooth muscle actin antibody, clone 1A4 – MP Biomedical #637931.

   Many companies market the same monoclonal antibody, clone 1A4, but not all preparations are equivalent. The antibody from MP Biomedical gives extremely strong staining, but the same antibody from Sigma has performed poorly in our hands.

9. Streptavidin, Alexa Fluor® 660 conjugate – Invitrogen #S21377.

10. Alexa Fluor® 350 F(ab')$_2$ goat antimouse IgG (H+L) – Invitrogen #A11068.

11. Fluoromount G – Electron Microscopy Sciences #17984-25.

3. Methods

There are several variations of this protocol, but the basic idea is common to all of them: detection of endothelial and smooth muscle cells in tissue sections using specific primary tags followed by visualization/amplification of those tags with secondary tags that can be visualized with a microscope. For this purpose, tissue sections can be from fixed paraffinized or fresh frozen tissue, and detection can be by fluorescent or histochemical staining. Opaque histochemical staining can result in a greater signal-to-background ratio and works well for staining of multiple markers as long as they do not colocalize. This can be done using a traditional biotin–avidin antibody approach, or a newer system from Biocare Medical that utilizes proprietary polymers also gives excellent results. Fluorescent staining is better if there is a chance that one staining pattern will overlap another. The fluorescent colors chosen will vary by need, but we have found that the blue fluorescence (e.g., DAPI) and the far red fluorescence (e.g., Cy5; CCD camera is required to detect) suffer from the least background autofluorescence in mouse cardiac tissue. The protocol provided here is for fluorescent detection in frozen tissues. The reagents suggested work on mouse cardiac tissue but should also work on tissue from other species.

Fluorescent detection of capillaries and arterioles

1. Thaw frozen sections.

2. Fix in 1.5% formaldehyde in PBS for 15 min.

3. Wash in PBS three times (e.g., 5 min, 5 min, 10 min).

4. Circle sections with PAP pen (marking pen that leaves hydrophobic residue).

5. Block/permeablize with staining buffer for 30 min at room temperature or overnight 4°C.

6. (If tissue is from mouse) Block with mouse-on-mouse block for 30 min.

7. Wash in PBS three times over 15 min.

8. Block in Avidin Block (from Dako Biotin Blocking System) for 10 min.

9. Wash in PBS three times over 15 min.

10. Block in Biotin Block (from Dako Biotin Blocking System) for 10 min.

11. Wash in PBS three times over 15 min.

12. Incubate in primary antibody and lectin as follows, diluted in staining buffer for 1 h at room temperature or overnight at 4°C.

    • Biotinylated Isolectin B_4 (1:100).

    • Mouse anti-α-smooth muscle actin (1:400).

13. Wash in staining buffer for 5 min, 5 min, and 10 min.

14. Incubate in fluorescent secondary antibody and avidin conjugates as follows, diluted in staining buffer for 1 h at room temperature or overnight at 4°C.

    • Streptavidin conjugated to Alexa 660 (1:100).

    • Goat antimouse $F(ab)_2$ conjugated to Alexa 350 (1:200).

15. Wash in staining buffer for 10 min.

16. Wash in PBS three times over 30 min.

17. Mount with Fluoromount and let harden overnight in the dark, then store in freezer.

18. View with fluorescence microscope using far red filter set (Cy5) for the Alexa 660 and the blue filter set (DAPI) for the Alexa 350; or other filter sets appropriate to the fluorophores of choice.

4. Notes

4.1. Histological Approaches Toward Assessment of Vascularity

You cannot indeed see more than one of my sections, or Circles, at a time; for you have no power to raise your eye out of the plane of Flatland; but you can at least see that, as I rise in Space, so my sections become smaller. See now, I will rise; and the effect upon your eye will be that my Circle will become smaller and smaller till it dwindles to a point and finally vanishes.

– from Flatland, by E.A. Abbott (1884)

In Flatland, a two-dimensional square living in a two-dimensional universe is visited by a three-dimensional sphere, but is only perceived by the square as whatever shape intersects with its two-dimensional existence. Thus, the unenlightened square only sees a circle that mysteriously grows and shrinks. This quote is useful and relevant in two ways. On the surface, it accurately describes how three-dimensional objects like capillaries are manifested in an effectively two-dimensional tissue section. More to the point of this chapter, it also points out that the observer's interpretation of what is seen in a tissue section is only as accurate as the observer's understanding of how the image was generated. One of the common mistakes made when quantitating vascularity is to determine that if a method of quantitation results in a number, that number must therefore mean something. This section describes several quantities that can be derived from histological tissue staining of blood vessels and shows how they can be either useful or misleading.

4.1.1. Counting Vessels

Probably the most straightforward concept in quantitating capillary vascularity is to "count capillaries," but what does this actually mean? Capillaries are a branched network, entirely interconnected, so how does one assign a number to this? By sectioning the tissue, we divide the capillary network into discrete units, vessel cross-sections that can be stained with antibodies or lectins that recognize endothelial cells and counted either manually or automatically by a computer. CD31 (PECAM-1) is the most common marker used for antibody staining. Antibodies against von Willebrand Factor are sometimes used but this staining does not always reveal all capillaries and sometimes only stains the large vessels, so it should be validated against other markers for new applications. Biotinylated BS-1 isolectin B_4 from *Griffonia simplicifolia* is a common lectin used in lieu of antibodies; while not officially endothelial-specific, it does an impressive job of staining all capillaries with minimal artifact and sometimes works under conditions that prevent good CD31 antibody staining. Note that staining directly with substrates for alkaline phosphatase, which is expressed in endothelial cells and used to be a popular method (2), has fallen out of favor because alkaline phosphatase expression is restricted to a minority of the vessels in the capillary bed (3). Arterioles can be counted in similar fashion after staining for smooth muscle markers like alpha smooth muscle actin (4). Counting objects has an advantage that a capillary will be counted the same way regardless of its obliqueness to the plane of the tissue slice, which affects the apparent size of its cross-section. However, this is still an oversimplification. If capillary counts from two tissue sections yield the same number of objects, but the average true diameter of the objects in one tissue is double that of the other, then a simple capillary count will show the two samples as being equivalent even though the vessels in one can move substantially more blood than the other.

Furthermore, it is important to determine whether the number of capillary cross-sections per unit area is a meaningful parameter in the specific case. For example, in skeletal muscle, edema associated with inflammation or injury can force muscle fibers apart significantly, decreasing the density of the capillaries per unit area but still allowing them to be associated with the same number of muscle fibers as in a nonedematous state (Fig. 1a, b). For this reason, it is common in skeletal muscle studies to report both the number of capillaries per mm² and the capillary/myocyte ratio (5).

In the myocardium, several similar considerations exist. Was the harvested heart stopped in diastole by KCl injection, or was it harvested without such intervention and allowed to come to a natural endpoint of systole (equivalent to *rigor mortis*); or was it put into formaldehyde while still beating and thus it "ground to a halt" at some undefined part of the beating cycle? Because the contracted myocardial wall at end-systole is thicker girth-wise than the extended myocardial wall at end-diastole, the capillaries may be farther apart at end-systole (Fig. 1c, d), so a measurement of capillaries per unit area will be more meaningful if all hearts have been stopped in the same controllable manner.

One further consideration is the way in which the objects were counted. The most foolproof approach is a manual count in a sufficiently large number of visual fields. The human eye can

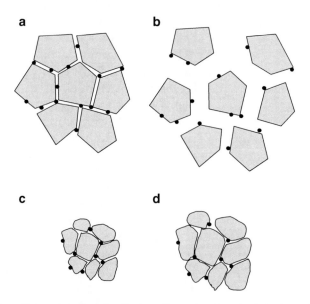

Fig. 1. Dependence of capillary density per unit area on edema in skeletal muscle (**a**, **b**) and on diastolic versus systolic dimensions of cardiomyocytes (**c**, **d**). In both cases, the number of capillaries per myocyte is unchanged but the absolute density per area will decrease as the capillaries are pushed farther apart from each other.

easily resolve stained vessel cross-sections from each other and from staining artifacts in ways that a computer may not be able to do. The main limitation of manual counts is that visual fields used for this approach are typically of high magnification, reducing the number of objects counted; and even if five such fields are counted for each sample (a common number in the literature), the total number of objects can remain statistically small and can lead to high variability. Manual counting can also take a fair amount of time to complete.

However, *computer-aided counting has an enormous pitfall* in all of these counting approaches, which is that if the ability of the computer to recognize the objects to be quantified is not accurate or reproducible, the ultimate number generated by the count can be meaningless. Automatic counting requires the user to first teach the software how to recognize the features to be counted, typically by recognizing certain shades of colors and/or intensities of vessel staining in a digital photo captured from the microscope. In the case of counting objects, typical errors can be introduced by (1) interpreting more than one object as a single object because they touch each other, (2) interpreting a single object as more than one object because of incomplete recognition of the stain (that is, counting two incomplete regions of stained object as two distinct stained objects), and (3) interpreting a staining artifact as an object. It is tempting to conclude that as long as the same errors are made in each sample, the errors will cancel each other out, but this is frequently not the case. No cryostat or microtome is perfect, and different tissue sections can be of slightly different thicknesses, resulting in different intensities of staining for reasons having nothing to do with biology. Even the staining itself can vary across a single section for a variety of reasons, so a computer setting that accurately recognizes objects in one part of the section may over- or undercount them in a different region. Therefore, despite a common tendency to set a threshold for computer recognition based on one excellent-looking section and using it blindly for all other sections, it is imperative to adjust (or at least confirm) the individual recognition parameters for each visual field to be counted, to ensure that the computer successfully recognizes what the user would have recognized manually. This clearly requires the investigator to be blinded to the identity of the tissue samples, and also can make the process of automatic counting closer to that of manual counting in terms of time.

It is easy to see from the above explanation how investigators can ascribe too much significance to the ability of a computer to analyze an image and generate a number.

4.1.2. Vessel Area Density   Another common approach to measuring vascularity is to determine how much area of the tissue section is accounted for by capillary staining, or what is sometimes called "vessel area density." This is a useful quantity because it distinguishes between large objects and small objects, rather than simply counting all objects as equivalent entities. It is primarily useful for capillary quantitation because arterioles span a vast range of wall thickness. It is also the easiest to determine digitally because it is relatively straightforward for a computer to count the number of pixels corresponding to capillary staining, assuming successful recognition (see below). This can be a weakness as well as a strength, however. If the difference in size between two vessel cross sections reflects a difference in size of the vessels themselves, then the results of the analysis are useful. However, a capillary that is completely perpendicular to the plane of the section shows up as a circle, while one that is oblique to the section will manifest itself as an oblong shape easily twice the length of the vessel diameter and potentially many times larger than that, depending on the angle (think about the Flatland quote above). Thus, if the difference between two vessel cross sections is actually reflective of a different orientation of the vessel relative to the plane of the section, then the computer is reporting a difference in vascularity that isn't really there. Here the counting of cross sections as objects holds a clear advantage, and simple quantitation of area presents serious caveats.

Let us examine this issue a bit more closely. If one is fortunate enough to work with a tissue that has obvious orientation, like skeletal muscle, in which the vast majority of capillaries are parallel to muscle fibers, then one can section the tissue at a relatively reproducible orientation and the variability between cross section angles is minimized. However, even in this case, all it takes is a slight difference in angle between one sectioned muscle and the next to introduce a measurable increase in diameter of almost all capillaries in that muscle, and a larger measured number. In the heart, different regions of the myocardium are at different angles and it is less straightforward to keep all samples at the same orientation. Moreover, the smaller the heart, the harder this is. For example, a block of pig myocardium can look reminiscent of skeletal muscle, with many cells at a constant orientation; whereas a mouse heart section can include numerous cardiomyocyte and capillary orientations in the same visual field. Therefore, capillaries in a mouse heart will be a mix of small circles and long strands, in which a single vessel at enough of an angle to the section can add an area density value equivalent to perhaps 20 or more single perpendicular vessels, skewing the count (Fig. 2). Again, it can be tempting to assume that any resulting variance in area measurements will be averaged out if one studies enough samples, and theoretically this is true, but practically it is a problem and at the very least will increase the standard deviations of the measurements to extents that potentially mask differences between samples.

Fig. 2. Capillaries in a 10 μm frozen section of a mouse heart, fluorescently stained with BS-1 isolectin B$_4$, which stains endothelial cells and is a good general capillary stain. Note that capillaries in some regions are small cross-sections while nearby regions consist of very long oblique capillaries, depending on the orientation of the particular region of myocardium.

At the worst, a variance that randomly favors one experimental group over another may detect a difference that does not actually exist.

Again, there are manual and digital approaches to determining area density. The digital approach is the easiest to conceptualize. One simply teaches the software to recognize capillary stain (see Subheading 4.1.4) and to count the number of positive pixels in the image. Here, it is important to establish a recognition standard that is as consistent as possible, because a slight difference in recognition threshold can lead to a huge difference in value. For example, a capillary cross section that is recognized as being 6 pixels in diameter will be reported with an area of $\pi r^2 = \pi \times 3^2 = 9\pi$ pixels, but if the recognition sensitivity is slightly higher and the computer recognizes one additional pixel of stain in all directions, then the reported area will be $\pi \times 4^2$ or 16π pixels – almost double the slightly less-sensitive example – for the same object. This can result from a slightly thicker section in which capillary cross sections have more mass to stain, or slight differences in the extent to which antibodies and rinse solutions have been aspirated off of the tissue during staining.

If a manual approach is desired, it is possible to determine area density by a somewhat indirect count compared to the digital counting of all pixels, but it is also less subject to artifacts and variability of staining and recognition efficiencies. In this approach, a constant grid pattern of some sort is superimposed against each image of the tissue stained for capillaries (software that uses multiple layers, like Adobe Photoshop or Improvision Openlab, is good for this purpose). Each overlap of a capillary with the grid is counted as one event (Fig. 3). A section with more stained

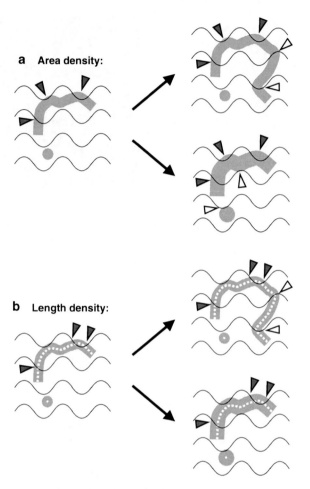

Fig. 3. Area density versus length density as quantifiable measures of vascularity. The *grey ribbon* represents a capillary segment nearly parallel with the tissue section, and the *grey circle* represents the cross section of a capillary that is perpendicular to the section. The *black sine wave* is the grid overlay, which would normally be in color. (**a**) In area density, any intersection with the grid is counted as one event (*dark arrowheads*). A comparable vessel of greater diameter or longer length both intersect the grid in more places (*white arrowheads*); therefore, area density is influenced by differences in girth and length of the vessels. (**b**) In length density, only an intersection of the imaginary vessel midline (*white dashed line*) with the grid is counted as an event. The comparable longer vessel still provides more intersection events but the thicker vessel does not; therefore; length density is influenced by length but not girth of the vessel.

capillary area in total will have more overlap events. Thus, the final count will not be expressed in meaningful units, but rather an index of area density; and comparisons between tissue sections with different extents of total vessel area will be easily discerned. The choice of grid pattern makes a large difference to the ultimate ease and speed of the counting. The pattern should not favor lines of any particular orientation because that can skew the probability of oriented objects like capillaries intersecting the pattern. A basic grid composed of repeating horizontal and vertical lines would not be very amenable to keeping track of what events have been counted. We find that a repeating sine wave pattern, with the distance between the waves determined partially by how dense the objects are being counted, gives a good representation of all orientations while still offering the investigator a single line to follow along at a time, recording overlap events as they occur.

4.1.3. Vessel Length Density

A useful variation of vessel area density is an approach that measures differences in cumulative vessel length without being influenced by differences in vessel area, referred to as vessel length density. In other words, if one simply wants to know how long the vessels are but does not care about their diameters, this is an appropriate quantity to measure. Simply put, length density measures the hypothetical result of taking every segment of capillary in the section and laying them end-to-end, and recording the final length. This can be useful, for example, if one desires to quantitate angiogenesis as linear growth in capillaries without concern for other processes that affect capillary diameter. Like area density, length density is primarily useful for capillaries rather than arterioles but could be used for the latter if the need arose.

Vessel length density is mostly appropriate for analysis situations in which the tissue is prepared in whole mount and intact vessels can be seen; for example, vasculature of mouse ear (6, 7), trachea (8), or pulmonary vasculature (9). In these cases, length density would be akin to tracing all vessels with a single line and recording the length of that line. In cardiac tissue in which most capillary cross sections show up as line segments due to oblique orientation to the section, length density can also yield meaningful numbers; while in sections in which capillaries are truly perpendicular and therefore the cross sections are dots or circles, length density is not as useful.

Length density can be determined manually using a simple variation of the grid overlay approach used for area density. The same grid pattern can be used, but in this case, rather than recording every overlap event in which any part of a vessel cross section touches the grid, only those overlap events are counted in which the pattern overlaps the imaginary midline of the vessel. In other words, if a capillary tube is envisioned with a central line running the length of the vessel, events are only counted if this line touches the pattern. As shown in Fig. 3, this can only be influenced by changes in linear dimension and is not affected by changes in girth.

A digital approach toward length density measurement is also possible but requires more sophisticated software, and does not always work. Certain software applications such as Image Processing Tool Kit plug-in for Adobe Photoshop (Reindeer Graphics)

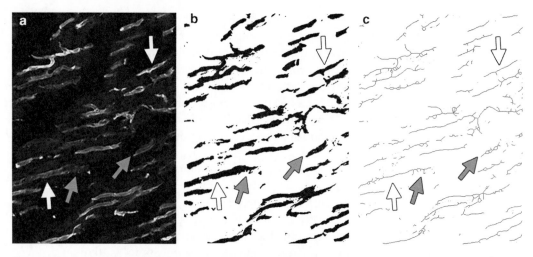

Fig. 4. Steps of "skeletonization" for length determination. Fluorescently-stained capillaries (**a**; *white against black*) are converted to a negative image (*black against white*) and put through a threshold conversion with a program such as Adobe Photoshop, converting the *grey* scale image to straight *black* and *white* (**b**). Alternatively, opaque histochemically stained capillaries (e.g., HRP staining) can be thresholded without a negative image. The resulting threshold image in (**b**) is subjected to a skeletonization conversion, found in several software applications including Image Processing Toolkit plug-in for Photoshop, which converts the 2-dimensional ribbons to 1-dimensional lines (**c**). Total line length can be easily measured by the total number of pixels in the skeleton image. Note regions in which conversion has been accurate (*white arrows*) and also regions in which aberrations in the texture of the original image cause short extraneous branches (*grey arrows*); the presence of enough of which can add considerable total line length to the image.

include "skeletonization" functions, in which a two-dimensional ribbon-shaped object is reduced to a one-dimensional line (Fig. 4). This line can then be measured in length by doing a simple count of pixels that comprise it. This strategy has been used quite successfully (10). However, objects of variable thickness or uneven edges can easily confuse skeletonization functions and can create substantial extra lengths of lines that are complete artifacts as shown in the figure. In the aforementioned study by Dutly et al. (10), an erosion and blurring features of Image Processing Tool Kit was used to remove spurious lines and pixels. Photoshop can also be used to manually remove lines that are obviously out of place, with special attention paid to thick vessels because they can show up in the skeletonization image as many lines perpendicular to the axis, doubling or even tripling the apparent length of that vessel. Once again, simply transforming an image by skeletonization without checking and fixing the result can lead to false confidence in meaningless results.

4.1.4. Important Considerations Regarding Training a Computer to Recognize Vessel Stain

The importance of validating the recognition of vessel stain by quantitation software cannot be overstated. We tend to implicitly trust whatever a computer tells us, but what is important to recognize is that the computer is not providing a magic analysis tool that takes any data fed into it regardless of quality and spits out meaningful numbers. It is merely providing a shortcut for the investigator to make measurements, and the investigator is responsible for ensuring that the data provided to the computer is solid, or else the investigator might as well read tea leaves instead. Therefore, as mentioned above, it is unwise to set a recognition parameter to accurately recognize vessel stain in a single section and then use that setting for the rest of the sections. Researchers sometimes do this in order to remove any bias from the measurements, but as long as the researcher is blinded to the identity of the samples, there can be no such bias. Indeed, by taking that approach, a slightly more robustly stained section can be overmeasured by that default setting, while a more weakly stained section can by the same token be undermeasured.

An example of this pitfall is illustrated in Fig. 5, which is based on a real example witnessed in a published paper that shall remain mercifully unreferenced. In this paper, the authors took the laudable extra step of showing examples of both their original capillary stains and the corresponding computer generated masks derived from automatic recognition of the stains. As shown in the simulation in the figure, a comparison of the stain and the recognition mask reveals that one image ("treatment") has been more faithfully recognized than the other ("negative control"), in which capillaries are slightly underrecognized, perhaps for reasons

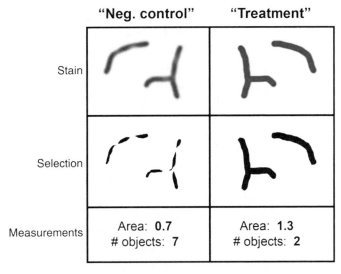

Fig. 5. Influence of staining quality on parameters of vascularity measurement. In this simulation based on a real example, the stain is slightly fainter in the "negative control" sample than in the "treatment" sample, potentially for reasons stemming from variability in tissue section thickness or individual rinsing conditions rather than a true biological response to the treatment. The computer thus selects smaller regions of stain in the control sample. This has the consequence of selecting more pixels in the treatment sample but more objects in the control sample, and the treatment could be claimed to have had a positive or negative effect on vascularity depending on which parameter was chosen to be presented.

such as discussed in Subheading 4.1.1. As a result, typical objects in the "treatment" image have more pixels recognized as stained, and the *capillary area density* is higher for the "treatment" image. However, the underrecognition of stain in the "control" image gives large single objects more of a tendency to be detected as three or four fragmented distinct objects, so the *capillary count* of the control sample will be higher than that of the treatment sample! Thus, this "treatment" could be reported to increase or decrease vascularity depending on which criterion is selected as the important one, and the conclusion of the same study could be success or failure. A number of published studies choose one approach or the other and simply call it angiogenesis, which underscores the peril of reading too much into quantitation of vascularity without understanding how the numbers were derived.

Assuming that the investigator will examine each image and tweak recognition settings to select vessel staining as accurately as possible, there are two main approaches toward software recognition of positive staining: color/intensity range and additive color selection ("magic wand" tool). Most software applications used for this purpose provide one or more sliders that allow the user to expand and contract ranges of color and intensity that will be selected in an image. Improvision's Openlab is an example of this category. As the user slides both ends of the selection range in and out, the area of the selection changes and the user can leave the sliders in the position that most accurately selects the stain and nothing more. This sometimes works very well, but it is common to choose ranges that faithfully select the desired stain in one region of the image but find that they over- or underselect in a different region, or in a different section. A more intuitive and potentially more effective selection approach uses the "magic wand" selection tool in Adobe Photoshop, also present in ImagePro (MediaCybernetics). By choosing the option within the application to make each successive click add selection to the previously selected pixels and setting the tolerance appropriately through trial and error, one can select a series of color shades representative of the stain. The ImagePro has a built-in quantitation feature, while modern versions of Photoshop have counting plug-ins that can be further enhanced with third party plug-ins like Image Processing Tool Kit (Reindeer Graphics). It is also acceptable to manually select or delete additional pixels to make the selection more accurately represent the stain. Again, as long as the investigator is blinded to the identity of the samples, the manual adjustment cannot be abused to support desired results.

4.1.5. Dealing with Incomplete Tissue

Each of these histological quantitation approaches provides a measure of vascularity per unit area. If the field of view in each image is constant, and the tissue is uninterrupted, then this is quite suitable. However, this frequently is not the case; either because the tissue photographed includes the edge of the tissue so part of the image is blank space, or because of rips or holes in the tissue. The missing tissue resulting from each example will be a region devoid of vessel count events that in no way reflects a lack of vascularity in tissue. For all of these approaches then, it is necessary to normalize the values obtained to the percentage of the field of view that actually contains tissue. This can be done digitally by having the software recognize anything that is either positive for vessel stain or exhibits any extent of tissue background fluorescence or coloration. Even if the vessel quantitation is done manually, a quick computer assessment of the percentage of field of view that contains tissue can furnish a correction factor with which the manual quantitation of vessels can be normalized. Alternatively, if the sine wave or grid overlay approach described in Subheading 4.1.2 is used, a rough approximation of the extent of tissue coverage can be manually obtained by counting the number of grid intersections or sine wave peaks and troughs that are in tissue and determine the percentage of those compared to the total number in the field of view (it is quicker to count the number that is NOT in tissue and subtract that percentage from 100).

4.1.6. The Peri-infarct Border Zone

One of the most common regions of cardiac tissue to be measured for vascularity is the peri-infarct border zone after a MI. There is a wide disparity in the approach that different investigators take toward the study of this region. Even the definition appears to change from research group to research group. When a region of myocardial tissue has died as a result of MI and remodeled into a scar, the border zone could be defined as the boundary between the two and the immediately surrounding tissue, or as the region of myocardium adjacent to the scar, which typically appears different from the rest of the myocardium and could be considered to still be at risk. Some investigators choose to find microscope fields at the border zone and compose pictures consisting of a constant ratio of scar and myocardium, e.g., two-third myocardium and one-third scar, and then count vessels indiscriminately throughout the entire visual field. If there is not a sharp delineation between the two, and there is a distinct border zone region consisting of a mixture of living cardiomyocytes and areas of scar, then this makes sense. However, the myocardium and the scar are like apples and oranges on the level of tissue, so if there is a clear separation between the two, it makes more biological sense to compare border zone myocardium under one condition with border zone myocardium under the other condition.

4.2. Functional Approaches Toward Assessment of Vascularity

The approaches discussed so far are all completely based on physical histological characteristics of the tissue to describe the number and cumulative length or area of blood vessels. However, it can be argued that this general approach falls short of evaluating the success of a provascularization treatment if it does not address the *functionality* of that vasculature. For the microvasculature, the most common functional parameter is the total amount of blood perfusion in the tissue.

4.2.1. Histological Endpoint Measurement of Vascular Perfusion

A classical approach toward histological assessment of functionality involves the injection into the upstream arterial circulation of 15 μm trackable plastic microspheres that are large enough that they become lodged in the capillary bed (11, 12). This does not cause physical problems because of the tiny number of microspheres injected relative to the total number of capillaries in the tissue. A higher number of lodged microspheres is indicative of a higher number of perfused capillaries. Originally, radioactive microspheres were used, which allowed quantitation of vascular perfusion by analysis of tissue samples in a scintillation counter. The modern version involves extremely bright fluorescent microspheres, which can yield quantitative results when the tissue has been dissolved in strong base to liberate the microspheres, which are in turn dissolved in an organic solvent so that the fluorescent dye contained therein can be analyzed by fluorimetry. An added benefit of the colored approach is that while the radioactive microspheres could only be injected once and studied at the end, the fluorescent microspheres can be injected at several time points over the course of the study, with each injected batch being of a different color. That way when the tissue is analyzed at the end, the different colors can give a read-out of what the functional perfusion was like at each of those timepoints.

The 15 μm microspheres can be quantitated at the level of the entire heart or muscle by measuring total fluorescence, but an alternative useful approach is to section the entire tissue thickly and either manually or digitally count the intact microspheres. While not giving a global assessment, this does allow regional patterns in the tissue to be recognized, which is especially useful if the procedure being studied is highly localized within the tissue.

Even though the 15 μm microsphere technique is considered a gold standard of functional tissue perfusion, a major caveat has been proposed by Ylä-Herttuala and colleagues (13), who have pointed out that some treatments that lead to substantial enlargement of capillaries can result in a large enough lumen to allow the microspheres to pass through unimpeded. A lack of lodged microspheres would significantly underestimate the perfusion using this method, and paradoxically, growth in cross-sectional size that increases perfusion would be measured as decreased perfusion. Once again, it is important to understand what is really occurring in the tissue being measured, rather than blindly using a technique and assuming that the data generated are meaningful.

While the 15 μm microsphere technique involves injection of a small number of microspheres into the circulation of living animals, there are several related approaches that are useful for visualizing capillaries that are in contact with the systemic circulation and can also be exploited for quantitation. Suspensions of much smaller fluorescent microspheres (0.2 μm) can be perfused through the vasculature of a small animal during euthanasia, either through directly into the left ventricle for studies of peripheral organs and muscles (14), or down the coronary artery for studies of the myocardium (15). The suspension will behave like a fluid because of the small size of the microspheres, but will still not leak through permeable vessels. The tissue can then be sectioned for microscopic analysis, or it can be dissolved away for fluorescent quantitation similar to that used for the larger microspheres. Similarly, a mixture of low-melt agarose and fluorescent microsphere suspension has been perfused through the vasculature such that the agarose prevents the fluorescent beads from diffusing away (9, 16). Total fluorescence can be measured in the tissue afterward for global measurements, and the fluorescent vessels can be analyzed by microscopy. One caveat with this approach is that slight changes in vessel diameters can result in large changes in vessel volumes and the amount of the tracer in the lumen. A potential variation of this – that would be influenced less by small changes in vessel diameters – may be to perform intravascular staining with a tail vein injection of biotinylated *Lysopersicon esculentum* lectin (8), which is typically used to visualize intact vessels by avidin/biotin staining after tissue harvest, but in this case the tissue could be homogenized and total biotinylated lectin could be quantitated by standard ELISA or blotting techniques.

4.2.2. Measurement of Vascular Perfusion in Living Animals

On a more gross physiological level, the standard approach to assessing perfusion of a tissue that does not involve euthanasia is to use laser Doppler perfusion measurement. A laser is shined on the surface of the tissue, and movements of blood in vessels within the shallow penetration depth, usually limited to skin, of the laser light are detected and measured. However, this is of little use for the heart because of its inaccessibility and its constant motion. Similarly, a variety of ultrasound-based Doppler techniques can be used in tumors and other stationary tissue. However, the beating of the heart also limits the utility of Doppler techniques in the heart to the study of inflow and outflow through heart valves, and motion of tissue.

MicroSPECT or microPET imaging can provide a gross approximation of perfusion in a small animal heart. This involves injection of radioactive tracers into the circulation and detection of signal in the perfused region of the heart. However, the resolution is relatively low for use in rodents.

In the last several years, ultrasound contrast techniques have evolved rapidly to the point of being able to evaluate perfusion in living hearts with a spatial resolution of as low as 30 μm depending on the technique being used. One approach involves the use of ultrasound contrast bubbles that behave like the large fluorescent microspheres described above, getting lodged in the capillary bed such that the level of contrast reveals the level of perfusion (17). The difference is that the result can be viewed noninvasively in the living animal, showing regional tissue differences such as that between an infarct scar and viable myocardium; and the contrast will gradually dissipate, allowing subsequent analyses at later time points. The other main approach is to use similar contrast bubbles that are in this case conjugated to antibodies against endothelial proteins, allowing literally the immunostaining of vessel lumens of living animals that can be visualized noninvasively by ultrasound (18). The ultrasound systems typically include software that can analyze the images from both of these techniques and quantitate the amount of contrast in regions of interest.

5. Conclusion

Histological staining of blood vessels and living animal imaging of vascular perfusion have the potential to offer a vast amount of useful information about the viability and stability of the tissue. However, as described in this chapter, the potential for misinterpretation of results is relatively high. A lack of understanding about how the quantitation approach really works can lead to minor errors or completely wrong conclusions. Conversely, having a strong grasp of these issues will ensure the most accurate conclusions about one's own measurements, as well as provide a way of evaluating the validity of reports in the literature.

References

1. Lu, Q. L., and Partridge, T. A. (1998) A new blocking method for application of murine monoclonal antibody to mouse tissue sections. *J Histochem Cytochem* **46**, 977–84.

2. Takeshita, S., Zheng, L. P., Brogi, E., Kearney, M., Pu, L. Q., Bunting, S., Ferrara, N., Symes, J. F., and Isner, J. M. (1994) Therapeutic angiogenesis. A single intraarterial bolus of vascular endothelial growth factor augments revascularization in a rabbit ischemic hind limb model. *J Clin Invest* **93**, 662–70.

3. Ushiki, T., and Abe, K. (1998) Identification of arterial and venous segments of blood vessels using alkaline phosphatase staining of ink/gelatin injected tissues. *Arch Histol Cytol* **61**, 215–9.

4. Springer, M. L., Ozawa, C. R., Banfi, A., Kraft, P. E., Ip, T. K., Brazelton, T. R., and Blau, H. M. (2003) Localized arteriole formation directly adjacent to the site of VEGF-induced angiogenesis in muscle. *Mol Ther* **7**, 441–9.

5. Tsurumi, Y., Takeshita, S., Chen, D., Kearney, M., Rossow, S. T., Passeri, J., Horowitz, J. R., Symes, J. F., and Isner, J. M. (1996) Direct intramuscular gene transfer of naked DNA encoding vascular endothelial growth factor augments collateral development and tissue perfusion. *Circulation* **94**, 3281–90.

6. Suri, C., McClain, J., Thurston, G., McDonald, D. M., Zhou, H., Oldmixon, E. H., Sato, T. N., and Yancopoulos, G. D.

(1998) Increased vascularization in mice overexpressing angiopoietin-1. *Science* **282**, 468–71.

7. Ozawa, C. R., Banfi, A., Glazer, N. L., Thurston, G., Springer, M. L., Kraft, P. E., McDonald, D. M., and Blau, H. M. (2004) Microenvironmental VEGF concentration, not total dose, determines a threshold between normal and aberrant angiogenesis. *J Clin Invest* **113**, 516–27.

8. Thurston, G., Baluk, P., Hirata, A., and McDonald, D. M. (1996) Permeability-related changes revealed at endothelial cell borders in inflamed venules by lectin binding. *Am J Physiol* **271**, H2547–62.

9. Zhao, Y. D., Courtman, D. W., Deng, Y., Kugathasan, L., Zhang, Q., and Stewart, D. J. (2005) Rescue of monocrotaline-induced pulmonary arterial hypertension using bone marrow-derived endothelial-like progenitor cells: efficacy of combined cell and eNOS gene therapy in established disease. *Circ Res* **96**, 442–50.

10. Dutly, A. E., Kugathasan, L., Trogadis, J. E., Keshavjee, S. H., Stewart, D. J., and Courtman, D. W. (2006) Fluorescent microangiography (FMA): an improved tool to visualize the pulmonary microvasculature. *Lab Invest* **86**, 409–16.

11. Jasper, M. S., McDermott, P., Gann, D. S., and Engeland, W. C. (1990) Measurement of blood flow to the adrenal capsule, cortex and medulla in dogs after hemorrhage by fluorescent microspheres. *J Auton Nerv Syst* **30**, 159–67.

12. Kowallik, P., Schulz, R., Guth, B. D., Schade, A., Paffhausen, W., Gross, R., and Heusch, G. (1991) Measurement of regional myocardial blood flow with multiple colored microspheres. *Circulation* **83**, 974–82.

13. Rissanen, T. T., Korpisalo, P., Markkanen, J. E., Liimatainen, T., Orden, M. R., Kholova, I., de Goede, A., Heikura, T., Grohn, O. H., and Ylä-Herttuala, S. (2005) Blood flow remodels growing vasculature during vascular endothelial growth factor gene therapy and determines between capillary arterialization and sprouting angiogenesis. *Circulation* **112**, 3937–46.

14. Springer, M. L., Ip, T. K., and Blau, H. M. (2000) Angiogenesis monitored by perfusion with a space-filling microbead suspension. *Mol Therapy* **1**, 82–7.

15. Christman, K. L., Fang, Q., Yee, M. S., Johnson, K. R., Sievers, R. E., and Lee, R. J. (2005) Enhanced neovasculature formation in ischemic myocardium following delivery of pleiotrophin plasmid in a biopolymer. *Biomaterials* **26**, 1139–44.

16. Leong-Poi, H., Kuliszewski, M. A., Lekas, M., Sibbald, M., Teichert-Kuliszewska, K., Klibanov, A. L., Stewart, D. J., and Lindner, J. R. (2007) Therapeutic arteriogenesis by ultrasound-mediated VEGF165 plasmid gene delivery to chronically ischemic skeletal muscle. *Circ Res* **101**, 295–303.

17. Kaufmann, B. A., Lankford, M., Behm, C. Z., French, B. A., Klibanov, A. L., Xu, Y., and Lindner, J. R. (2007) High-resolution myocardial perfusion imaging in mice with high-frequency echocardiographic detection of a depot contrast agent. *J Am Soc Echocardiogr* **20**, 136–43.

18. Lyshchik, A., Fleischer, A. C., Huamani, J., Hallahan, D. E., Brissova, M., and Gore, J. C. (2007) Molecular imaging of vascular endothelial growth factor receptor 2 expression using targeted contrast-enhanced high-frequency ultrasonography. *J Ultrasound Med* **26**, 1575–86.

Part IV

Measures of Cell Trafficking

Chapter 11

Superparamagnetic Iron Oxide Labeling of Stem Cells for MRI Tracking and Delivery in Cardiovascular Disease

Dorota A. Kedziorek and Dara L. Kraitchman

Abstract

In the mid-1980s, iron oxide nanoparticles were developed as contrast agents for diagnostic imaging. In the last two decades, established methods to label cells with superparamagnetic iron oxides (SPIOs) have been developed to aid in targeted delivery and tracking of stem cell therapies. The surge in cellular therapy clinical trials for cardiovascular applications has seen a similar rise in the number of preclinical animal studies of SPIO-labeled stem cells in an effort to understand the mechanisms of cardiovascular regenerative therapy and stem cell biodistribution. The adoption of a limited number of methods of direct labeling of stem cells with SPIOs is due in large part to the desire to rapidly translate these techniques to clinical trials. In this review, we will outline the most commonly adopted methods for iron oxide labeling of stem cells for cardiovascular applications and describe strategies for magnetic resonance imaging (MRI) of magnetically labeled cells in the heart.

Key words: Magnetic resonance imaging (MRI), Stem cells, Superparamagnetic iron oxide (SPIO), Cellular labeling, Cell imaging, Transfection, Electroporation

1. Introduction

Magnetic resonance imaging (MRI) provides many desirable features for cell tracking and delivery. MRI offers the interactivity of X-ray interventional techniques without exposing the patient or cells to ionizing radiation. Moreover, the high spatial resolution and exquisite soft tissue detail of MRI are superior to X-ray cardiac interventional methods, which can only provide information about the lumen of the heart or vessels in combination with iodinated contrast agents. In addition, MRI allows noninvasive, serial imaging for dynamic tracking of cell migration and engraftment (1–12).

Randall J. Lee (ed.), *Stem Cells for Myocardial Regeneration: Methods and Protocols*, Methods in Molecular Biology, vol. 660, DOI 10.1007/978-1-60761-705-1_11, © Springer Science+Business Media, LLC 2010

Although there are many MRI cellular labeling methods, direct cellular labeling with superparamagnetic iron oxide (SPIO) contrast agents have been mostly widely used for many reasons (13–15). The lack of unique surface markers for stem cells that are retained with stem cell differentiation down a cardiac lineage has limited the feasibility of receptor-based labeling methods. In addition, direct cell labeling methods are relatively simple, fast, and inexpensive. Several clinically approved formulations of SPIO-based contrast agents are available that have been used for cell labeling in a variety of diseases. Toxicity of these agents is low, since the SPIO nanoparticles that are released from dying cells can be degraded in the normal iron recycling pathways. Compared to gadolinium-based contrast agents, SPIOs become more effective upon cell internalization due to particle clustering and, thereby, create large "blooming" hypointensities on standard clinical MRI scanners. While SPIOs are not internalized natively by nonphagocytic cells, simple methods to induce internalization and uptake have been developed and tested in a variety of stem cells (2, 4, 11, 16–18). One of the most common methods is "magnetofection" – a method where transfection agents (TAs) are used to coat SPIOs to encourage endocytosis of the SPIO–TA complex (13, 14). Concentrations of 2–10 pg iron/cell can be achieved after 12–48 h incubation in vitro (13). After exogenous labeling of stem cells, the SPIOs are stably maintained in endosomes and have been imaged for several months after delivery to the heart (3, 5, 9, 18).

Magnetoelectroporation (MEP) is another common method of SPIO cellular labeling. Magnetoelectroporation uses small pulsed voltages to encourage endocytosis of SPIOs (19). No transfection agents are needed, which may aid in more rapid clinical translation. In addition, millions of cells can be labeled in seconds using magnetoelectroporation, which may be important in certain cell lines that are altered by culturing in vivo. Furthermore, for cardiac cellular delivery, magnetoelectroporation may prove to be the method of choice where cellular delivery cannot be delayed by 24–48 h after an acute cardiac event.

Both magnetofection and magnetoelectroporation can be used to label cells with a variety of contrast agents. A recent study performing a head-to-head comparison of magnetofection and MEP demonstrated preserved cell viability and proliferation in embryonic stem cells by both techniques (20). However, cardiac differentiation of embryonic stem cells was most attenuated by MEP and iron uptake was greatest with magnetofection (20). Detailed methods using these techniques to label stem cells for cardiovascular applications using SPIO contrast agents will be given in this review.

2. Materials

2.1. Cell Culture and Preparation

1. Stem cell media, suitable for cell origin and requirements, for example, MEM alpha supplemented with 10% fetal bovine serum (FBS, HyClone, Logan, UT, USA) and 1% antibiotic/antimycotic containing penicillin, streptomycin, and amphotericin B (Gibco/Invitrogen, Grand Island, NY, USA) for mesenchymal stem cells.

2. 10 mM Phosphate Buffered Saline (PBS) (1×), pH = 7.4.

3. Trypsin (0.5 g/L) with ethylenediaminetetraacetic acid (EDTA 0.2 g/L) (Gibco/Invitrogen, Grand Island, NY, USA), warmed to 37°C in a water bath.

2.2. Labeling with Transfection Agents

TAs are highly charged molecules that will form complexes with iron oxide particles through electrostatic interactions. There are several classes of these agents, but the most convenient and commonly utilized labeling methods are those based on commercially available TAs including dendrimers, such as Superfect, poly-L-lysine (PLL), Lipofectamin, and FUGENE. Iron oxide magnetic nanoparticles are used in conjunction with TAs to label cells and can be distinguished primarily based on the size of the nanoparticles. For brevity, we list several formulations that are approved or under development by major pharmaceutical concerns.

1. Iron oxide contrast agents:

    (a) Commercially available ferumoxides are Feridex (Berlex Laboratories Inc., Wayne, NJ, USA) or Endorem (Guerbet SA, Paris, France). Ferumoxide stock solution contains 11.2 mgFe/mL with particles approximately 80–150 nm in diameter (21). Ferumoxide stock solution should be stored at 4°C; do not freeze! Feridex is an FDA-approved liver contrast agent since 1996. In Europe, this compound is registered under name Endorem. Both agents contain a dextran coating to minimize clumping.

    (b) Ferucarbotran (Resovist, Bayer Schering Pharma AG, Berlin, Germany) is an SPIO composed of a colloidal solution of iron oxide nanoparticles coated with carboxydextran. It is currently used for the detection and characterization of focal liver tumor lesions and approved for clinical use in the European, Australian, and Japanese markets (see Note 1).

    (c) Ferumoxtran (Sinerem, Guerbet SA, Paris, France or Combidex, AMAG Pharmaceuticals Inc., Cambridge, MA, USA) is a member of the ultrasmall superparamagnetic iron oxide (USPIO) class of contrast agents with a median diameter <50 nm. Due to the smaller diameter, these particles will not be filtered by the reticuloendothelial system as quickly as SPIOs when injected intravenously. Thus, they tend to accumulate in lymph nodes and are used to distinguish normal from metastatic nodes (see Note 2).

2. Transfection agents:

(a) PLL (PLL, Sigma, St Louis, MO, USA) as the polyamine PLL hydrobromide with molecular weight of 388,199 Daltons (catalog number P-1524) is the most commonly used TA. A stock solution of PLL at concentration of 1.5 mg/mL should be stored in −20°C.

(b) Protamine sulfate (American Pharmaceuticals Partner, Schaumburg, IL, USA), which is a drug used clinically to reverse the effects of heparin therapy, is another commonly used TA. It is available in bottles at a concentration of 10 mg/mL.

2.3. Magnetoelectroporation

1. Ferumoxide stock solution (see Subheading 2.2).

2. Electroporation cuvettes, 0.4 mm gap (Gene Pulser BioRad, Hercules, CA). It is important to deliver electrical pulses with the proper field strength and duration. The exact pulse delivery will be dependent on the type of cells that will be labeled. Mammalian cells typically require field strengths up to 6.15 kV/cm, which can be obtained using the 0.4 cm cuvette.

3. BTX electroporation system (Harvard Apparatus, Holliston, MA, see Fig. 1).

4. Culture media and 10 mM PBS as in Subheading 2.1.

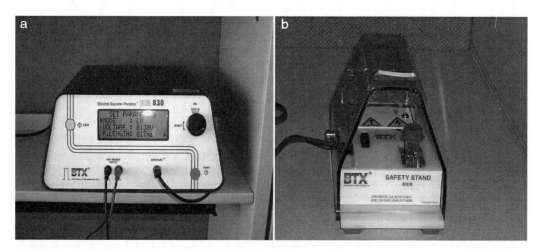

Fig. 1. BTX electroporation system (**a**) that can be used for magnetoelectroporation. This system may be operated by a switch (**b**) or foot pedal (not shown).

2.4. Magnetic Resonance Imaging	1. Clinical MRI scanner equipped with surface coils to image the heart.
	2. MRI-compatible ECG leads and monitoring equipment.

---

3. Methods

There are two commonly used iron oxide labeling methods: (1) labeling with transfection agents, for example, PLL (22) or protamine sulfate (21); and (2) Magnetoelectroporation (19, 23). Prior to labeling, the cells must be prepared in a clean, appropriate environment. The cell can be frozen and thawed immediately prior to labeling, but viability is improved if the cells are brought back into culture before labeling.

3.1. Cell Culture and Preparation

1. When the stem cells approach confluence in a culture dish/flask, remove the old media and wash the monolayer once or twice with PBS.

2. Remove PBS with a pipette and add a minimal volume (~1 mL for a T-75 flask) of prewarmed trypsin.

3. Incubate the cells in trypsin for at 37°C in humidified, enriched in 5% CO_2 air then check microscopically to determine whether the monolayer of cells is lifting off the culture dish after ~2-3 minutes.

4. When single cell suspension has been obtained, add ~10 mL of complete media and transfer the cell suspension to a sterile 15 mL conical tube and spin the cells on tabletop centrifuge (~$600 \times g$ for 10 min).

5. Discard the supernatant carefully, so as not to disturb the cell pellet and resuspend the cells in complete media for counting.

6. Once the number and concentration of cells has been determined, reseed the cells. For example, MSCs typically will be reseeded into a fresh T-75 flask at a concentration of ~2×10^5 cells/mL.

3.2. Labeling with Transfection Agent

Each combination of TA and (U)SPIOs should be carefully titrated and optimized, since too low concentrations may not lead to sufficient cellular uptake, whereas too high concentrations may induce precipitates (see Note 3) or may be cytotoxic

3.2.1. Labeling with (U) SPIOs–PLL Complexes

1. Allow the cells to grow to 80–90% confluence of the culture dish surface area.

2. While maintaining clean suitable environment for cell culture, prepare complete media appropriate for the stem cell type at

the standard volume needed for regular cell growth. Add (U) SPIOs to media solution to obtain final concentration of 25 µg Fe/mL (13, 14, 24) and mix the solution for few seconds with intermittent hand shaking to obtain homogenous solution.

3. Add PLL to achieve final concentration of 375 ng/mL (2, 24). When using Sigma stock solution (as described in Subheading 2.2), add 0.25 µL of this stock per 1 mL of the ferumoxide/media mixture. Mix the solution shaking gently and incubate for 30–60 min with occasional gentle hand mixing.

4. After the (U)SPIO–PLL media solution is fully mixed, discard the old media from the cells, wash the monolayer with PBS, and add the media containing (U)SPIO–PLL complex to the culture flask. Incubate the cells with this solution overnight, for example, approximately 16–18 h at 37°C in air enriched with 5% CO_2 (see Note 4).

5. After overnight incubation, remove the media containing (U) SPIO–PLL complexes, rinse the cells with warm PBS, trypsinize, and collect for counting and administration. The lack of species specificity of SPIO–PLL labeling is shown for MSCs in Fig. 2.

3.2.2. Cell Labeling with (U)SPIOs: Protamine Sulfate Complexes

1. Dilute protamine sulfate in distilled water to the concentration of 1 mg/mL.

2. Combine (U)SPIOs with appropriate serum-free media based on the cell type to obtain a concentration of 100 µg Fe/mL. For example, add 9 µL of ferumoxide formulation for every 1 mL of media for MSCs.

3. Add protamine sulfate to the (U)SPIO solution to obtain its concentration of 4.5–6 µg/mL (25–27) and hand shake the solution periodically over 5–10 min.

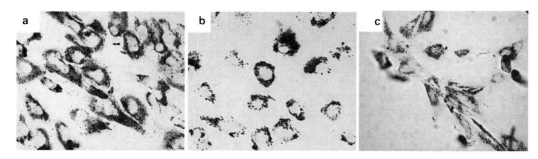

Fig. 2. Mesenchymal stem cells (MSCs) using ferumoxides–PLL. Cells were labeled for 48 h (**a**, **b**) or 24 h (**c**) with 25 mg Fe/ml Feridex and 375 ng/ml PLL. DAB-enhanced Prussian blue stain of labeled human (**a**), porcine (**b**), and canine (**c**) MSCs show an efficient intracellular uptake of particles into endosomes that is nonspecific across species. (Adapted from Bulte and Kraitchman (5) with permission.)

4. After 5–10 min, add an equal volume of standard cell culture media with a double concentration of serum to create final ferumoxide concentration of 50 µg/mL.

5. Replace old media in cell culture with newly created media with (U)SPIO–protamine sulfate complexes and incubate with cells overnight.

6. After overnight incubation, remove the media containing (U)SPIO–protamine sulfate complexes, rinse the cells with warm PBS, trypsinize, and collect for counting and administration.

3.3. Magnetoelectro-poration

1. Remove media and wash the cells with PBS.

2. Trypsinize and count the cells. After counting, spin the cells on tabletop centrifuge ($\sim$600 $\times g$ for 10 min for MSCs) and wash with PBS.

3. Resuspend the cells in 10 mM sterile PBS at the density of 1.5×10^6 cells/mL (see Note 5) and transfer to sterile electroporation cuvette(s). While cell suspensions <1 mL/cuvette may be used, care must be taken to ensure that the cuvettes' electrodes are entirely covered by the cell suspension. For example, using the BTX apparatus and 0.4 mm gap electrooration cuvettes, the total volume of cell suspension mixed with (U)SPIOs cannot be smaller than 700 µL.

4. Add (U)SPIOs to obtain a final concentration (after mixing with cell suspension) of 2,000 µg Fe/mL. For example, mix 130 µL of ferumoxides with 600 µL of cell suspension to obtain a final electroporation mix volume equal to 730 µL (see Note 6).

5. Using BTX electroporation system, electroporate cells using the following conditions: 50 V pulse strength; 5 ms pulse duration; and 20 pulses in intervals of 100 ms (see Note 7).

6. Leave the cuvettes in the holder for 1 min, transfer to ice, and let them to rest for 5 min to allow for membrane recovery.

7. Remove the small top layer of foam and transfer cells to 50 mL conical tube containing culture media. Leave the tube with cells on ice for at least 15–20 min (see Note 8).

8. Wash the cells twice with PBS. Spin the cells in media on tabletop centrifuge ($\sim$600 $\times g$ for 10 min).

9. Pipette off the supernatant, resuspend MSCs in fresh, sterile 10 mM PBS, and spin again. Repeat steps 8 and 9 and then proceed to step 10.

10. Discard the supernatant and resuspend cell pellet in 1 mL (or other desired amount) of PBS. Count the cells and dilute to final concentration for administration.

3.4. Cardiac Magnetic Resonance Imaging

(U)SPIO-labeled stem cells can be delivered in several ways including direct visualization during open-chest procedures, intracoronary administration using conventional angiographic catheters, and transmyocardially using specialized catheters for delivery of therapeutics to the myocardium. High spatial resolution T2*-weighted images will depict the labeled cells as hypointensities. Because these hypointensities can be hard to distinguish from other hypointensities, such as calcified plaque or metallic objects like stents, off-resonant imaging techniques have been developed to portray the magnetic susceptibilities from iron-labeled cells as hyperintense signal (28–32).

3.4.1. T2*-Weighted Imaging

1. For high-resolution imaging, images are acquired over multiple cardiac cycles using ECG-gating. Motion artifacts from breathing are suppressed using either navigator echo techniques or breath-holding.

2. While T2*-weighting can be obtained using several imaging techniques, gradient echo imaging with an extended echo time (TE) that does not degrade cardiac images appears to provide the best compromise in image quality on clinical scanners (see Fig. 3) (2).

3. Typical gradient echo imaging parameters are: 6 ms repetition time (TR); 1.6 ms TE; 20° flip angle; 512 × 512 image matrix; 5–8 mm slice thickness (ST): 32 kHz bandwidth (BW); and 2–4 number of signal averages (NSA). Images are acquired in the standard short or long axis planes to cover the extent of the left ventricle.

3.4.2. Off-Resonance Imaging

There are several types of off-resonance imaging techniques that have been used to image (U)SPIOs. One method uses a spectral excitation in combination with spin echo imaging, which is probably not well-suited for cardiac applications (29). Another

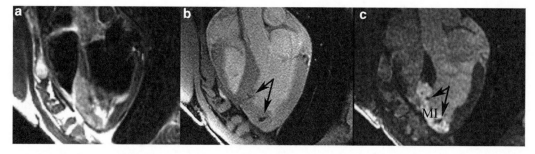

Fig. 3. Representative hypointensities (*arrows*) from ferumoxide–PLL-labeled mesenchymal stem cell in long-axis MRIs of an infarcted swine using fast spin echo image (**a**), fast gradient echo (**b**), and delayed contrast-enhanced MRI (**c**), which demonstrates infarcted myocardium as hyperintensities. Labeled-MSCs were delivered transmyocardially using a specialized MRI-compatible injection catheter. (Adapted from Kraitchman et al. (2) with permission.)

method, GRadient echo Acquisition for Superparamagnetic particles/suscePtibility (GRASP), modifies the refocusing pulses to create positive contrast from iron-labeled cells (30, 33). Background suppression using this technique is excellent. A third method, Inversion-Recovery with ON-resonant water suppression (IRON), uses frequency-selective prepulses to suppress the water signal leaving positive contrast from iron-labeled cells (31). While not providing as much background suppression as the GRASP method, IRON MRI provides the flexibility to be combined with either spin echo or gradient echo techniques as well as two-dimensional single plane or three-dimensional volume acquisitions. Recently, it has been demonstrated that these off-resonance imaging techniques will not benefit from field strengths >4.7 T (34). Thus, these techniques are ideal for use on currently available clinical scanners.

1. As with T2*-weighted imaging, ECG-gating and respiratory gating or breath holds are used to suspend cardiac and respiratory motion, respectively.

2. Typical imaging parameters for three dimensional fast spin echo IRON imaging at 3 T are: 2 ms TR; 11.6 ms TE; 24 echo train length (ETL): 11.6 ms interecho spacing; 170 Hz bandwidth water suppression; 95° iron saturation pulse. In the heart, fat suppression is recommended. Depending on the number of (U)SPIO-labeled cells per voxel and the image resolution, the off-resonant positive contrast will appear as hyperintense areas surrounding the cells and with a typical dipole appearance (see Fig. 4). The volume of the hyperintensities can be measured to determine a relative concentration of labeled cells.

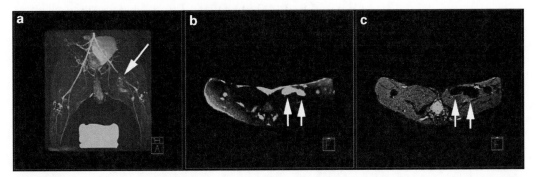

Fig. 4. A maximum intensity projection of a 3D T2-prepared MR angiogram (**a**) acquired on a 3 T MR clinical scanner in a rabbit model of peripheral arterial disease. Mesenchymal stem cells (MSCs) that were labeled using magnetoelectroporation with ferumoxides were injected into the medial thigh. Two injection sites imaged immediately after injection are shown with IRON MRI (*arrows*, **b**) and conventional gradient echo imaging (*arrows*, **c**). (Adapted from Kraitchman and Bulte (32) with permission.)

4. Notes

1. Because ferucarbotrans are smaller than ferumoxides, the ratio of SPIO to cationic transfection agents may need to be adjusted. This may explain why unpublished data have suggested that aggregates using ferucarbotrans are more common. However, Politi and coworkers were able to more efficiently label neural precursor cells with ferucarbotrans than ferumoxides or USPIOs (35).

2. While USPIO labeling has been performed in cardiovascular applications (36), a few reports suggest that cell labeling with SPIOs is more efficient than USPIOs (35, 37, 38).

3. Cahill et al. filtered ferumoxide–PLL complexes using a 0.2 μm mesh to reduce aggregates that would otherwise form during incubation with muscle stem cell transplants in media (39).

4. If overnight labeling of cells is not possible, cells can be also efficiently labeled in 2 h by using serum-free media with ferumoxide–PLL complexes formation and cell nourishment (22).

5. If cell density per cuvette exceeds 5×10^6, cell clumping can occur during magnetoelectroporation procedure. Thus, using less than 2×10^6 cells per cuvette is recommended to avoid cell clumping (19).

6. Iron uptake will be determined in part by of the type of cells. Cells with more cytoplasmic volume can be labeled with a larger amount of SPIOs (see Fig. 5). In addition, increasing the SPIO concentration can enhance intracellular iron uptake during magnetoelectroporation. Walczak and coworkers have demonstrated a correlation between concentrations ranging from 250 to 2,000 of Fe μg/mL and cellular iron uptake (see Fig. 6) (19).

7. Augmenting the number of pulses with a lower voltage (40 mV) can also enhance intracellular iron uptake by magnetoelectroporation (19). Iron uptake quantity by rat MSCs sufficient for MR imaging (more than 2 pg of iron per cell) were achieved even with the small amount of 250 of Fe μg/mL (19). However, labeling with small amounts of iron may not permit cell tracking if the cells proliferate and, thereby, dilute the label.

8. Our studies in MSCs have found that incubation in media on ice after magnetoelectroporation increases cell viability. We would speculate that this "rest" period allows the membrane to recover prior to exposure to a harsh environment in vivo.

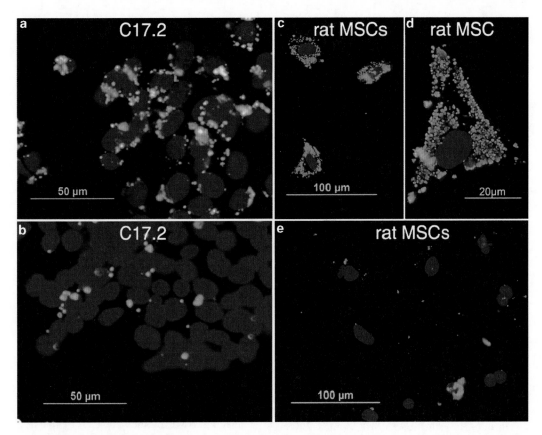

Fig. 5. Iron uptake by MEP-treated cells. C17.2 mouse neural stem cells (**a**, **b**) and rat mesenchymal stem cells (**c–e**) were incubated with ferumoxides (2 mg Fe/mL) with (**a**, **c**, **d**) and without (**b**, **e**) magnetoelectroporation (MEP). Only MEP-treated cells show significant ferumoxide uptake as assessed by antidextran immunofluorescent staining (*green*). The higher magnification in (**d**) demonstrates ferumoxide-containing clusters with a measured diameter of 830 ± 350 nm. (From Walczak et al. (19) with permission.)

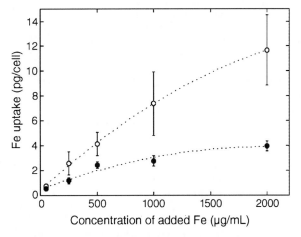

Fig. 6. Iron content of MEP-labeled rat MSCs (*open circles*) and C17.2 cells (*closed circles*). The amount of cellular iron uptake increases with the added amount of ferumoxides. Rat MSCs, which have a larger cytoplasmic volume, contain more iron than smaller C17.2 cells. (From Walczak et al. (19) with permission.)

References

1. Weissleder R, Cheng HC, Bogdanova A, Bogdanov A, Jr. Magnetically labeled cells can be detected by MR imaging. J Magn Reson Imaging 1997;7(1):258–63.

2. Kraitchman DL, Heldman AW, Atalar E, et al. In vivo magnetic resonance imaging of mesenchymal stem cells in myocardial infarction. Circulation 2003;107(18):2290–3.

3. Hill JM, Dick AJ, Raman VK, et al. Serial cardiac magnetic resonance imaging of injected mesenchymal stem cells. Circulation 2003; 108(8):1009–14.

4. Garot J, Unterseeh T, Teiger E, et al. Magnetic resonance imaging of targeted catheter-based implantation of myogenic precursor cells into infarcted left ventricular myocardium. J Am Coll Cardiol 2003;41(10):1841–6.

5. Bulte JW, Kraitchman DL. Monitoring cell therapy using iron oxide MR contrast agents. Curr Pharm Biotechnol 2004;5(6):567–84.

6. Rickers C, Gallegos R, Seethamraju RT, et al. Applications of magnetic resonance imaging for cardiac stem cell therapy. J Interv Cardiol 2004;17(1):37–46.

7. Kustermann E, Roell W, Breitbach M, et al. Stem cell implantation in ischemic mouse heart: a high-resolution magnetic resonance imaging investigation. NMR Biomed 2005; 18(6):362–70.

8. de Vries IJ, Lesterhuis WJ, Barentsz JO, et al. Magnetic resonance tracking of dendritic cells in melanoma patients for monitoring of cellular therapy. Nat Biotechnol 2005;23(11): 1407–13.

9. Stuckey DJ, Carr CA, Martin-Rendon E, et al. Iron particles for noninvasive monitoring of bone marrow stromal cell engraftment into, and isolation of viable engrafted donor cells from, the heart. Stem Cells 2006;24(8): 1968–75.

10. Amado LC, Schuleri KH, Saliaris AP, et al. Multimodality noninvasive imaging demonstrates in vivo cardiac regeneration after mesenchymal stem cell therapy. J Am Coll Cardiol 2006;48(10):2116–24.

11. Ebert SN, Taylor DG, Nguyen HL, et al. Noninvasive tracking of cardiac embryonic stem cells in vivo using magnetic resonance imaging techniques. Stem Cells 2007;25(11): 2936–44.

12. Arai T, Kofidis T, Bulte JW, et al. Dual in vivo magnetic resonance evaluation of magnetically labeled mouse embryonic stem cells and cardiac function at 1.5 t. Magn Reson Med 2006;55(1):203–9.

13. Frank JA, Miller BR, Arbab AS, et al. Clinically applicable labeling of mammalian and stem cells by combining superparamagnetic iron oxides and transfection agents. Radiology 2003;228:480–7.

14. Frank JA, Zywicke H, Jordan EK, et al. Magnetic intracellular labeling of mammalian cells by combining (FDA-approved) superparamagnetic iron oxide MR contrast agents and commonly used transfection agents. Acad Radiol 2002;9:S484–S7.

15. Kalish H, Arbab AS, Miller BR, et al. Combination of transfection agents and magnetic resonance contrast agents for cellular imaging: relationship between relaxivities, electrostatic forces, and chemical composition. Magn Reson Med 2003;50(2):275–82.

16. Cahill KS, Germain S, Byrne BJ, Walter GA. Non-invasive analysis of myoblast transplants in rodent cardiac muscle. Int J Cardiovasc Imaging 2004;20(6):593–8.

17. Tallheden T, Nannmark U, Lorentzon M, et al. In vivo MR imaging of magnetically labeled human embryonic stem cells. Life Sci 2006;79(10):999–1006.

18. Bulte JW, Kraitchman DL. Iron oxide MR contrast agents for molecular and cellular imaging. NMR Biomed 2004;17(7):484–99.

19. Walczak P, Kedziorek D, Gilad AA, Lin S, Bulte JW. Instant MR labeling of stem cells using magnetoelectroporation. Magn Reson Med 2005;54(4):769–74.

20. Suzuki Y, Zhang S, Kundu P, Yeung AC, Robbins RC, Yang PC. In vitro comparison of the biological effects of three transfection methods for magnetically labeling mouse embryonic stem cells with ferumoxides. Magn Reson Med 2007;57(6):1173–9.

21. Arbab AS, Yocum GT, Kalish H, et al. Efficient magnetic cell labeling with protamine sulfate complexed to ferumoxides for cellular MRI. Blood 2004;104(4):1217–23.

22. Bulte JW, Arbab AS, Douglas T, Frank JA. Preparation of magnetically labeled cells for cell tracking by magnetic resonance imaging. Methods Enzymol 2004;386:275–99.

23. Walczak P, Ruiz-Cabello J, Kedziorek DA, et al. Magnetoelectroporation: improved labeling of neural stem cells and leukocytes for cellular magnetic resonance imaging using a single FDA-approved agent. Nanomedicine 2006;2(2):89–94.

24. Kostura L, Kraitchman DL, Mackay AM, Pittenger MF, Bulte JW. Feridex labeling of mesenchymal stem cells inhibits chondrogenesis

but not adipogenesis or osteogenesis. NMR Biomed 2004;17(7):513–7.

25. Janic B, Iskander AS, Rad AM, Soltanian-Zadeh H, Arbab AS. Effects of ferumoxides-protamine sulfate labeling on immunomodulatory characteristics of macrophage-like THP-1 cells. PLoS One 2008;3(6):e2499.

26. Rad AM, Janic B, Iskander AS, Soltanian-Zadeh H, Arbab AS. Measurement of quantity of iron in magnetically labeled cells: comparison among different UV/VIS spectrometric methods. Biotechniques 2007;43(5):627–8, 30, 32 passim.

27. Pawelczyk E, Arbab AS, Pandit S, Hu E, Frank JA. Expression of transferrin receptor and ferritin following ferumoxides-protamine sulfate labeling of cells: implications for cellular magnetic resonance imaging. NMR Biomed 2006;19(5):581–92.

28. Seppenwoolde JH, Viergever MA, Bakker CJ. Passive tracking exploiting local signal conservation: the white marker phenomenon. Magn Reson Med 2003;50(4):784–90.

29. Cunningham CH, Arai T, Yang PC, McConnell MV, Pauly JM, Conolly SM. Positive contrast magnetic resonance imaging of cells labeled with magnetic nanoparticles. Magn Reson Med 2005;53(5):999–1005.

30. Mani V, Saebo KC, Itskovich V, Samber DD, Fayad ZA. GRadient echo Acquisition for Superparamagnetic particles with Positive contrast (GRASP): Sequence characterization in membrane and glass superparamagnetic iron oxide phantoms at 1.5 T and 3 T. Magn Reson Med 2006;55:126–35.

31. Stuber M, Gilson WD, Schär M, et al. Positive contrast visualization of iron oxide-labeled stem cells using inversion recovery with ON-resonant water suppression (IRON). Magn Reson Med 2007;58:1072–7.

32. Kraitchman DL, Bulte JW. Imaging of stem cells using MRI. Basic Res Cardiol 2008;103(2):105–13.

33. Mani V, Briley-Saebo KC, Hyafil F, Itskovich V, Fayad ZA. Positive magnetic resonance signal enhancement from ferritin using a GRASP (GRE acquisition for superparamagnetic particles) sequence: ex vivo and in vivo study. J Cardiovasc Magn Reson 2006;8(1):49–50.

34. Farrar CT, Dai G, Novikov M, et al. Impact of field strength and iron oxide nanoparticle concentration on the linearity and diagnostic accuracy of off-resonance imaging. NMR Biomed 2008;21(5):453–63.

35. Politi LS, Bacigaluppi M, Brambilla E, et al. Magnetic-resonance-based tracking and quantification of intravenously injected neural stem cell accumulation in the brains of mice with experimental multiple sclerosis. Stem Cells 2007;25(10):2583–92.

36. Nelson GN, Roh JD, Mirensky TL, et al. Initial evaluation of the use of USPIO cell labeling and noninvasive MR monitoring of human tissue-engineered vascular grafts in vivo. FASEB J 2008;22(11):3888–95.

37. Oude Engberink RD, van der Pol SM, Dopp EA, de Vries HE, Blezer EL. Comparison of SPIO and USPIO for in vitro labeling of human monocytes: MR detection and cell function. Radiology 2007;243(2):467–74.

38. Sun R, Dittrich J, Le-Huu M, et al. Physical and biological characterization of superparamagnetic iron oxide- and ultrasmall superparamagnetic iron oxide-labeled cells: a comparison. Invest Radiol 2005;40(8):504–13.

39. Cahill KS, Gaidosh G, Huard J, Silver X, Byrne BJ, Walter GA. Noninvasive monitoring and tracking of muscle stem cell transplants. Transplantation 2004;78(11):1626–33.

Chapter 12

Embryonic Stem Cell Biology: Insights from Molecular Imaging

Karim Sallam and Joseph C. Wu

Abstract

Embryonic stem (ES) cells have therapeutic potential in disorders of cellular loss such as myocardial infarction, type I diabetes and neurodegenerative disorders. ES cell biology in living subjects was largely poorly understood until incorporation of molecular imaging into the field. Reporter gene imaging works by integrating a reporter gene into ES cells and using a reporter probe to induce a signal detectable by normal imaging modalities. Reporter gene imaging allows for longitudinal tracking of ES cells within the same host for a prolonged period of time. This has advantages over postmortem immunohistochemistry and traditional imaging modalities. The advantages include expression of reporter gene is limited to viable cells, expression is conserved between generations of dividing cells, and expression can be linked to a specific population of cells. These advantages were especially useful in studying a dynamic cell population such as ES cells and proved useful in elucidating the biology of ES cells. Reporter gene imaging identified poor integration of differentiated ES cells transplanted into host tissue as well as delayed donor cell death as reasons for poor long-term survival in vivo. This imaging technology also confirmed that ES cells indeed have immunogenic properties that factor into cell survival and differentiation. Finally, reporter gene imaging improved our understanding of the neoplastic risk of undifferentiated ES cells in forming teratomas. Despite such advances, much remains to be understood about ES cell biology to translate this technology to the bedside, and reporter gene imaging will certainly play a key role in formulating this understanding.

Key words: Embryonic stem cells, Molecular imaging, Reporter gene imaging, Immunologic response, Teratoma, Cell transplantation

1 Introduction

Embryonic stem (ES) cell therapy has promising therapeutic implications for disorders of cellular loss such as coronary artery disease, diabetes mellitus, and neurodegenerative disease (1–3). While there has been considerable enthusiasm for stem cell

Randall J. Lee (ed.), *Stem Cells for Myocardial Regeneration: Methods and Protocols*, Methods in Molecular Biology, vol. 660, DOI 10.1007/978-1-60761-705-1_12, © Springer Science+Business Media, LLC 2010

therapy, the initial excitement has been tempered after multiple barriers met in the lab, which make it unlikely that ES cells will be clinically applicable in the immediate future. These challenges are directly related to stem cell differentiation and survival, immunogenic response, and potential for tumorgenicity (4). Studies using traditional postmortem immunohistochemistry have proven insufficient for understanding the process of ES cell biology. The recent development of novel molecular imaging modalities has revolutionized the ability to track stem cells in vivo, making breakthroughs in solving these challenges possible for the first time.

2. Molecular Imaging

There are two main types of imaging modalities: probe-based imaging and reporter gene-based imaging. Probe-based imaging includes conventional magnetic resonance imaging (MRI), positron emission tomography (PET), and single photon emission computed tomography (SPECT). Reporter gene-based imaging is based on linking a reporter gene to a promoter of interest. Receptor binding or enzymatic interaction between the reporter gene product (i.e., reporter protein) and reporter probe then generates an appreciable signal that can be measured.

Probe-based imaging involves stem cells assimilating a given probe that produces a quantifiable signal, which is then imaged with a specialized detector. For instance, iron oxide is the labeling probe used most commonly in MRI of stem cells, and it usually has to be coupled to a bioactive compound (e.g., lipofectamine) to promote uptake into the ES cells. After cells are delivered to the target organs, the animal can be imaged. MRI can identify the location of the iron labeled cells with high spatial resolution for a relatively long period of time (4–16 weeks). However, there are several major drawbacks with this modality. First, a relatively high number of cells are required to produce a detectable signal (5). Second, iron oxide particles can be transferred to nearby cells but *not* necessarily to daughter cells, making it less than ideal in tracking a rapidly dividing population such as that of ES cells. Finally, iron oxide can remain latent in dead tissue or can be taken up by macrophages so its presence does not automatically imply cell viability (6).

Radionuclide imaging utilizes Technetium-99 (^{99}Tc), Indium-111 (^{111}In), or Iodine-123 (^{123}I) to track the fate of stem cells. For instance, SPECT can be used to detect high photon emissions from probe containing cells. Similarly, PET detects photon signals from 2-fluoro-2-deoxy-D-glucose ([^{18}F]-FDG), which is a bioactive glucose molecule taken up by pretreated cells. With poorer spatial resolution but significantly higher sensitivity than MRI, PET has proven useful in tracking stem cells in vivo in patients with acute myocardial infarction (7). Because of their short half-lives, conventional probes for PET and SPECT are only detectable in a matter of hours or days, much quicker than the several weeks typically seen with iron-oxide probes used for MRI. As in the case of MRI, radionuclide probes may diffuse into neighboring cells without reliable transfer to daughter cells. It therefore has the same drawback in that signal enhancement does not necessarily imply viability, making it difficult to conclude that the cells of interest are actually viable.

To overcome many of the limitations of probe-based imaging, reporter gene imaging technology has been developed. Reporter gene imaging utilizes the same specialized detection (MRI, SPECT, and PET) as probe-based imaging, with the difference being that the reporter gene is first inserted into the cell's genome. Cells are transduced or transfected with a nonviral or viral delivery system containing a reporter gene linked to promoter of interest. Once incorporated into the genome, expression of the protein is dependent on the expression of the promoter or enhancer element, which can be inducible, constitutive, or tissue specific. When subsequently injected with a reporter probe, the reporter protein-reporter probe interaction produces a signal detectable by one of the aforementioned methods.

Reporter gene imaging has multiple advantages over probe-based imaging techniques. First, the reporter gene can be integrated into the genome, linking its expression with the viability of the cell of interest. In addition, by ensuring stable integration of the reporter gene into the cell's DNA, stable transmission to daughter cells is achievable. Likewise, by linking the reporter gene to a tissue-specific promoter, the emission signals can indicate the selective differentiation of stem cells into specific cell types. Finally, by inserting multiple reporter genes into the cell, multimodality imaging can be utilized to view the same cell (8). The need for genetic manipulation of the cells is the major drawback to this modality. It is worth noting, however, that this manipulation has *not* affected the viability or differentiation of ES cells (9, 10). The most commonly used reporter genes are firefly luciferase (Fluc) and herpes simplex virus-thymidine kinase (HSV-tk). Fluc-based systems utilize a bioluminescence imaging (BLI) system. Fluc catalyzes D-luciferin, which produces low energy photons (2–3 ev) that can be captured by a charge-coupled device (CCD) camera. On the other hand, HSV-tk phosphorylates 9-(4-18F-fluoro-3-[hydroxymethyl]butyl)guanine ($[^{18}F]$-FHBG), which produces high energy photons that can be detected by PET. The added advantage of this technique is that HSV-tk can serve as a suicide gene when ganciclovir is administered (8, 11). In addition, monomeric red fluorescence protein (mRFP) and enhanced green fluorescence protein (eGFP) reporter genes can be used in selection of cells through fluorescence-activated cell sorting (FACS).

3. Stem Cell Therapy

3.1. Stem Cell Differentiation and Survival

Murine ES cells were first isolated almost 30 years ago, whereas human ES cells were isolated only about 10 years ago (12–14). Since then, considerable effort has been undertaken to harness the pluripotent potential of those cells to differentiate into any cell type, principally by inducing ES cell differentiation into several cell types in vitro (1–3). However, early in vivo experiments in rodents showed inconsistent cell survival and variable patterns of differentiation (15–17). It became clear that a more robust method of tracking ES cell in vivo would be essential to understand the differentiation behavior and survival of transplanted stem cells. Postmortem immunohistochemistry provides only a single snapshot of cell fate at a given time point. On the other hand, probe-based imaging utilizing iron oxide labeling and [111]Indium is limited by many of the aforementioned drawbacks such as probe diffusion and inability to discern cell viability. For these and other reasons, reporter gene-based imaging has emerged as a versatile modality that has a track record of providing important insights into stem cell differentiation and engraftment (6).

The notion that injection of ES cell-derived endothelial cells (ESC-ECs) can improve cardiac function was tested by Li and colleagues (15). Here a double fusion reporter gene self-inactivating lentiviral vector driving Fluc and mRFP was used to stably transduce mouse ES cells. Coronary artery ligation was used as a model for inducing myocardial injury, followed by injection of reporter gene-labeled stem cells or PBS. BLI showed a massive drop in cell survival over the first week followed by a small persistent signal up to 8 weeks. Postmortem double staining for CD31 (endothelial marker) and mRFP confirmed the presence of ESC-ECs in the peri-infarct region. Interestingly, this corresponded to a small but statistically significant improvement in left ventricular fractional shortening (FS) of the heart at week 8. The ability to monitor cell kinetics in this case made it possible to conclude that the majority of injected ESC-ECs do not survive following transplantation.

Experiments involving human ES cell-derived cardiomyocytes (hESC-CMs) utilized similar methods to investigate their survival rate following transplantation (1). hESC-CMs were isolated from beating embryoid bodies followed by Percoll density gradient separation. These cells were injected into the infarcted myocardium. BLI revealed a sharp drop-off in signals in the first 2–3 weeks followed by steady levels up to the 8-week period (Fig. 1). Postmortem immunohistochemistry showed small areas of differentiated myocytes but no true integration into the host myocardium. Nevertheless, the mouse hearts showed a noteworthy transient improvement in FS around 8 weeks, which disappointingly did

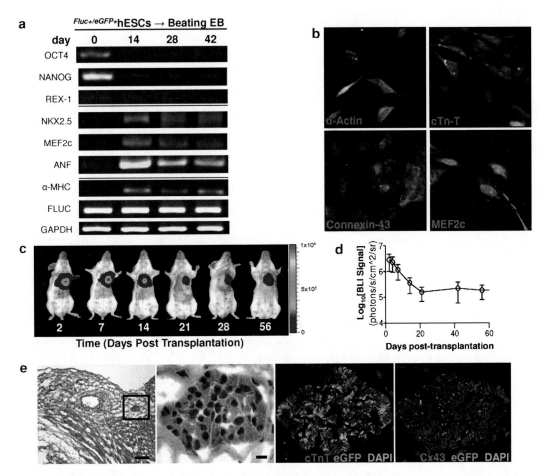

Fig. 1. Survival and fate of *Fluc+/eGFP+*hESC-CMs in vivo. (**a**) RT-PCR analysis of various hESC and cardiac specific markers revealed no significant differences between *Fluc+/eGFP+*hESCs and control nontransduced hESCs, other than the presence of Fluc. (**b**) *Fluc+/eGFP+*hESC-CMs express cardiac specific markers such as α-actin, troponin-T, connexin-43, and MEF2C (all in *red*) and GFP (*green*, scale bars = 50 μm). (**c**) A representative animal imaged for 2 months following transplantation of one million *Fluc+/eGFP+*hESC-CMs into the heart. (**d**) In vivo bioluminescence imaging (BLI) signal measured from animals in which *Fluc+/eGFP+*hESC-CMs were transplanted into the ischemic hearts ($n = 15$). Signal activity falls drastically within the first 3 weeks of transplantation and remains stable thereafter, with no evidence of tumorigenesis (*left*). From 21 days posttransplantation onwards, BLI signal is reduced to <10% of the signal obtained at 2 days posttransplantation. (**e**) Histopathological evaluation of hearts following *Fluc+/eGFP+*hESC-CM delivery. H&E staining (*left panels*) demonstrates cluster of cells within the infarcted region of the heart (scale bars = 200, 20 μm for low and high magnification images, respectively). GFP-positive cells within this cluster also express cardiac troponin-T (*red*, near *right panel*) and connexin-43 (*red*, far *right panel*). Scale bars = 20 μm. Adapted with permission from Cao et al. (1).

not persist by 16 weeks. In this case, the imaging data complemented the immunohistochemical and functional results, making possible a much better understanding of the relationship between cell survival and functional restoration.

Directed eGFP expression can be utilized beyond cell tracking, including the identification and isolation of cells in vitro in preparation for transplantation. For example, Duan et al. investigated the differentiation of human ES cells into hepatocytes by utilizing a lentiviral vector containing eGFP linked to a hepatic specific promoter (α-1-antitrypsin) (18). Expression of eGFP allowed for laser microdissection in vitro under a fluorescent microscope. Molecular analysis confirmed that the majority of isolated cells expressed hepatocyte specific genes. In addition, the investigators injected transduced human ES cells into livers of SCID/NOD mice. Within the first week of injection, fluorescence signals could be detected. This is an example of in vivo tracking of a specific cell line by linking the reporter gene to a tissue-specific promoter.

In summary, understanding stem cell survival in vivo remains a complex task. It is clear that random sampling of animals at different time points for postmortem analysis provide a limited snapshot instead of a longitudinal view of cellular survival, migration, and proliferation. Molecular imaging approach has allowed us to better appreciate the complexity of ES cell survival in vivo. Furthermore, these results demonstrate that ES cell derivatives (especially ESC-ECs and ESC-CMs) often fail to survive long term. This process is also seen in adult stem cells such as bone marrow mononuclear cells, skeletal myoblasts, mesenchymal stem cells, and fetal cardiomyocytes (19, 20). The reasons for donor cell death remain poorly understood, and active research is under way at many laboratories.

3.2. Immunogenicity

In its infancy, ES cell therapy was heralded as an immune-safe treatment (21, 22). This made ES cell therapy a particularly attractive alternative to orthotopic organ transplantation, which relies heavily on immunosuppressive pharmacotherapy. The immunity advantage was widely believed and not challenged until the past few years, when our understanding of ES cells began to improve. The debate continues to some extent because of the divergent reported results regarding the survival of transplanted ES cells, ranging from full rejection to no immune response. These studies involve a heterogeneous group of cells and target organs as well as varying methods of assessing immune response and survival, making comparisons across studies difficult.

Intuitively, ES cells must have mechanisms of evading the immune response, given that fetal tissues are composed partially of "foreign" paternal origin. Li et al. showed that ES cells could inhibit the proliferation of allogeneic T cells in vitro, but the full mechanism of immune system evasion in vivo remains unclear (21). Further complicating the analysis is the fact that mouse ES cells do not have the same immunologic behavior as their human counterparts, making direct extrapolations an unsafe wager (4).

Molecular imaging has contributed significantly to establishing that last fact. Swijnenburg et al. sought to evaluate this process by taking advantage of reporter gene imaging techniques (23). Mouse ES cells were stably transduced with an inactivated lentiviral vector containing Fluc-eGFP driven by a constitutive ubiquitin promoter. Allogeneic and syngeneic ES cells were injected into leg muscles. By day 14, ES cell survival was significantly higher in the syngeneic group. Postmortem analysis showed teratoma formation in the syngeneic group, but not in the allogeneic group. The presence of teratoma formation suggests that mouse ES cells in the syngeneic group survived and differentiated while those in the allogeneic group did not. Furthermore, FACS analysis showed that the transplant regions contained inflammatory cells with the allogeneic transplant group showing higher proportions of CD3+ and CD8+ cells. This suggests that cytotoxic T-cell pathway is specifically involved in the rejection of mouse ES cells. Finally, the investigators transplanted another group of animals *previously sensitized* to mouse ES cells and showed that those mice experienced much faster BLI signal loss. This implies a more rapid rejection due to donor-specific adaptive immune response. Overall, these experiments were the first to highlight several important facts. First, mouse ES cells are not immuno-privileged in vivo and can elicit both innate and adaptive immune responses resulting in rejection of transplanted cells. Second, by evading the immune response through the use of syngeneic model, ES cells can persist and differentiate (although the differentiation in this experiment progressed to teratoma, which is a different topic that should be addressed separately).

The behavior of human ES cells was investigated in a similar set of experiments by utilizing a similar Fluc-eGFP double fusion reporter gene driven by a constitutive human ubiquitin promoter (24). BLI was used for serial monitoring of transplanted human ES cells into immunocompetent and immunodeficient (SCID/NOD) mice. BLI signals in the immunocompetent mice disappeared by day 7–10, whereas in immunodeficient mice the signal intensified and persisted until animals were sacrificed at day 42. Furthermore, re-transplantation of ES cells in the contralateral leg showed an accelerated loss of signal in the immunocompetent group (Fig. 2). This confirmed that human ES cells, in an xenogeneic model, are not immune-privileged and indeed can trigger an innate and an adaptive immune response. In addition, the investigators evaluated the effect of multiple immunosuppressive regimens on the survival of transplanted human ES cells in immunocompetent mice. Once again, BLI allowed for serial monitoring of the signal intensity of ES cells over time, which showed improved survival with the combination therapy of tacrolimus and sirolimus, albeit not as robust as that seen in SCID mice.

In summary, the initial disappointment from the realization of the immunogenic potential of human ES cells should be replaced by gratification that this immune response can partly explain the conflicting survival and teratoma formation results in prior experiments. Reporter gene imaging was instrumental in

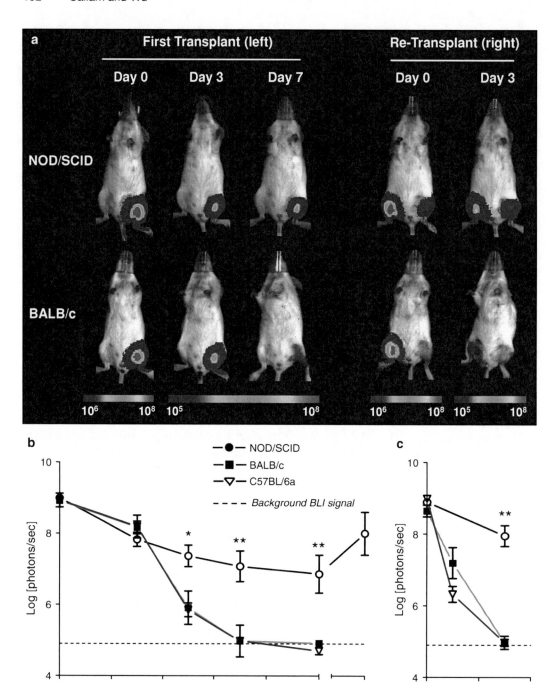

Fig. 2. In vivo visualization of $^{Fluc+/eGFP+}$hESC survival. (**a**) Representative BLI images of $^{Fluc+/eGFP+}$hESC transplanted animals show a rapid decrease in BLI signal in immunocompetent animals (BALB/c), as opposed to immunodeficient (NOD/SCID) mice, reaching background levels at posttransplant day 7. Accelerated BLI signal loss in BALB/c animals was seen following repeated hESC transplantations into the contralateral gastrocnemius muscle. Color scale bar values are in photons/s/cm²/sr. Graphical representation of longitudinal BLI after (**b**) primary and (**c**) secondary hESC transplantation into immunodeficient (NOD/SCID, $n=5$) and two immunocompetent (BALB/c and C57Bl/6a, $n=5$ per group) mouse strains. Note that in NOD/SCID animals, starting at posttransplant day 10, BLI intensity increases progressively, suggesting hESC proliferation. $*P < 0.05$, $**P < 0.01$. Adapted with permission from Swijnenburg et al. (24).

providing a method of noninvasive serial monitoring of stem cell kinetics. It has several advantages compared to probe-based imaging techniques, such as: (1) its ability to provide a fairly accurate surrogate of cell number and viability without interfering with cell function; (2) a fluorescent property that allows for postmortem tissue and FACS analysis; and (3) the ability to make quantitative comparisons with other images over time and across different animals. More research is needed to understand the immunology of human ES cells. Multimodality imaging techniques, with their obvious advantages, will certainly play a critical role in furthering our understanding of this topic.

3.3. Teratoma Formation

Early experiments injecting ES cells in vivo into animals made it clear that the pluripotent differentiation potential of ES cells carries a risk of teratoma formation (4). The biology of teratoma formation remains imperfectly understood, but advances in molecular imaging have provided valuable insights into the basics of this aberrant differentiation. For instance, the purity of injected cell population has been proposed as a risk factor for teratoma formation, with a higher risk associated with a larger number of undifferentiated ES cells. However, precise details of the level of purity remain largely unknown. To investigate the minimum number of undifferentiated cells that can lead to teratoma formation, a recent study injected 1, 10, 100, 1,000 and 10,000 mouse ES cells stably expressing Fluc-eGFP into the subcutaneous dorsal region of nude mice (25). After 3 months, the animals injected with 1,000 ES cells developed teratomas, whereas no teratomas were observed by BLI or postmortem immunohistochemistry in animals injected with 1, 10, or 100 cells. This would suggest the safety margin is between 100 and 1,000 for mouse ES cells. A similar experiment was repeated with human ES cells utilizing the double fusion Fluc-eGFP reporter gene (1). With human ES cells, teratoma formation was observed at the 100,000-cell impurity level but not with the 10,000- or 1,000-cell impurity level.

In a follow-up study, mouse ES cells stably expressing double fusion (Fluc-GFP) and triple fusion (Fluc-mRFP-HSV-tk) were injected subcutaneously into the left and right shoulders of immunodeficient mice, respectively (26). Similar to the findings of the prior experiment, BLI and PET showed increased signals in both injection sites consistent with teratoma development by 2–3 weeks. Subsequently, animals were infused with pharmacological doses of ganciclovir, which can inhibit DNA synthesis and ablate the specific cells possessing the HSV-tk. After administration of ganciclovir, cell signals in the *right* shoulder (contains HSV-tk) showed a complete return to baseline levels of BLI and PET signal (Fig. 3). Furthermore, postmortem analysis confirmed tumor necrosis in those areas, whereas persistent growth was seen in the left shoulder (lacks HSV-tk). In this case, reporter gene imaging

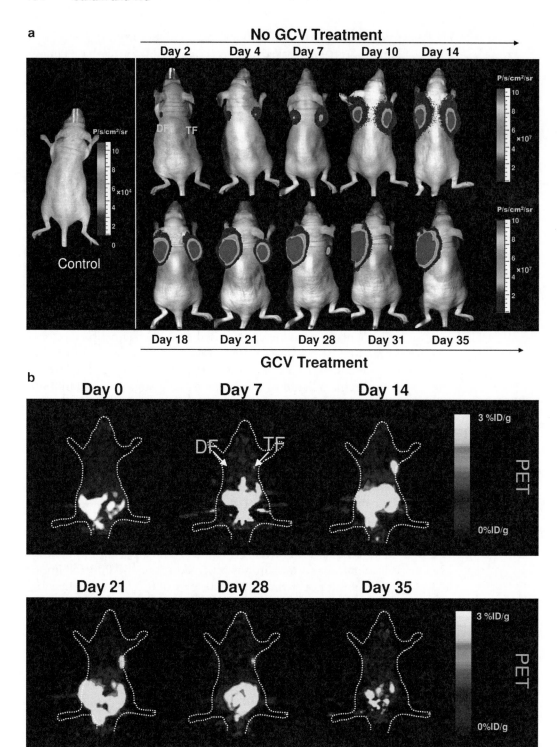

Ganciclovir treatment started at day 15

not only monitored teratoma development, but also served as part of the therapeutic intervention against cellular misbehavior.

In summary, advanced molecular techniques have allowed us to derive multiple new conclusions regarding teratoma formation. This set of experiments show the clear advantage of being able to survey the entire host, leading to the realization of the full body risk of teratoma formation. In addition, BLI allows us to identify 100–1,000 of mouse ES cells and 10,000–100,000 of human ES cells as the purity margin for teratoma formation. Future research should be directed at improving the purity level of differentiated cells, identifying injection target organs that may be less vulnerable to teratoma formation, and finding interventions to inhibit or ablate teratoma formation in vivo. Reporter gene imaging will continue to be an important tool in these endeavors because of its ability to provide sensitive measures of cell fate in vivo.

4. Conclusion

ES cells are characterized by their pluripotent capacity and unlimited self-renewal ability, making them ideal candidates for cell replacement therapies. Unfortunately, this great potential comes with complex biology and has remained elusive. Although ES cells remain far from clinical trials, it is safe to say that our understanding of their behavior has improved markedly over the past few years. Reporter gene imaging techniques share considerable credit for elucidating many aspects of our understanding of ES cell differentiation, immunology, and tumorgenic potential. The prospective translation of ES cells from the bench to the bedside will certainly rely on reporter gene imaging to clarify cell fate and discern the effects of different treatments, among other important roles.

Fig. 3. Noninvasive imaging of mouse ES cell survival, proliferation, and teratoma formation. (**a**) One million ES cells carrying Fluc-eGFP (ES-DF) were injected subcutaneously into the left shoulder and one million ES cells carrying Fluc-mRFP-HSV-ttk (ES-TF) cells into the *right shoulder*. Serial imaging of the same representative animal over 2 weeks showed progressive increase in imaging signals (*top row*). To prevent teratoma formation, ganciclovir treatment was started from week 2 to week 5. At week 5, ES-TF cells carrying the HSV-ttk reporter-suicide gene showed background bioluminescence imaging while ES-DF cells lacking HSV-ttk showed dramatic increases in signal activity. (**b**) PET imaging of selective ablation of ES-TF teratoma formation at the *right shoulder* from week 2 to week 5. The *left shoulder* showed background signal because there was no uptake of the PET reporter probe [^{18}F]-FHBG by ES-DF cells. Activity is present in the gut and bladder region due to the natural excretion route of [^{18}F]-FHBG. Adapted with permission from Cao et al. (26).

5. Materials

5.1. Virus Production/ Transfection

1. 293T cell (human embryonic kidney fibroblasts) growth medium: minimal essential media (MEM) supplemented with 10% fetal bovine serum (FBS) and 1% penicillin (100 μg/ml)/streptomycin (292 μg/ml).

2. 2.5 M $CaCl_2$, stock solution (sterilize through a 0.45-μm filter and store at –20°C).

3. 2× BES (N,N-Bis(2-hydroxyethyl)-2-aminoethanesulfonic acid)-buffered saline (BBS): 50 mM BES (pH 6.95), 280 mM NaCl, and 1.5 mM Na_2HPO_4. Sterilize through a 0.45-μm filter and store at –20°C. The pH can be adjusted with HCl at room temperature.

4. HIV-1 packaging vector (pCMVΔR8.2) and vesicular stomatitis virus G glycoprotein-pseudotyped envelop vector (pMD.G).

5. Self-inactivating lentiviral vector containing triple fusion construct (pFUG-TF). (Details of construction of triple fusion reporter gene are described by Ghambhir et al. (27)).

6. 4:1 mixture of ketamine and xylazine.

5.2. Bioluminescence Imaging

1. In Vivo Imaging System 100 (IVIS 100, Caliper Life Sciences, Hopkinton, MA).

2. D-Lucefrin solution at a concentration of 45 mg/ml.

3. 4:1 mixture of ketamine and xylazine.

5.3. PET Imaging

1. MicroPet scanner (R4 Concorde, Knoxville, TN).

2. [^{18}F]-fluoro-3-(hydroxymethyl) butyl guanine ([^{18}F]-FHBG).

3. Acquisition Sinogram and Image Processing (ASIPro) software (Siemens Medical Solutions USA).

4. AMIDE software for multimodality image analysis (28).

5. 4:1 mixture of ketamine and xylazine.

6. Methods

6.1. Virus Production

1. Passage and plate 293 T and seed at approximately 5×10^5 cells/plate.

2. Mix 15 μg pFUG-TF (containing the triple fusion construct), 10 μg HIV-1 packaging vector (pCMVΔR8.2) and 5 μg vesicular stomatitis virus G glycoprotein-pseudotyped envelop vector (pMD.G) with 0.25M $CaCl_2$.

3. Add 0.5 ml 2× BBS.

4. Incubate at room temperature for 30 min.

5. Add the calcium phosphate-DNA solution one drop at a time to a plate of 293T cells, mixing well between each drop.

6. Incubate the cells for 48–72 h, changing the media at 16–24 h.

7. Aspirate the supernatant containing the lentiviral vector and centrifuge at 3,000 rpm for 5 min.

8. Purify the lentiviral vector by passing the solution through a 0.45-µm filter.

9. Centrifuge the purified solution with a SW 29 rotor at $50,000 \times g$ for 2 h.

10. Dissolve the viral sediment in 100 µl of serum-free medium.

11. Transduction efficiency of p-FUG-TF by target cells can be assessed using fluorescence microscopy and transfected cells can be isolated using FACS analysis. Store the viral medium at –70°C.

6.2. Viral Transfection

1. Always wear safety goggles when handling viral elements. Viral stock should be thawed at room temperature.

2. Thawed solution can directly be added to target cells at a concentration of approximately 10^6 cells.

3. Refresh culture medium at 12 and 24 h.

4. Animal is anesthetized for injection using 4:1 mixture of ketamine and xylazine (2 µl/gm body weight) intraperitoneally.

5. After animal is anesthetized, transfected cells can be injected into target organ in vivo.

6.3. Bioluminescence Imaging

1. Animal is injected with 4:1 mixture of ketamine and xylazine (2 µl/gm body weight) intraperitoneally.

2. 5–10 min later, inject D-Lucefrin into the animal at concentration of 125 mg/kg for imaging of deep organs and 375 mg/kg for superficial imaging.

3. Place the animals in a light-tight chamber and obtain baseline gray-scale body-surface images.

4. Image animals for up to 30 min using 1 min acquisition intervals.

5. Quantify bioluminescence in units of photons per second per centimeter square per steridian ($p/s/cm^2/sr$) as described by Cao et al. (1).

6.4. PET Imaging

1. Animal is injected with 4:1 mixture of ketamine and xylazine (2 µl/gm body weight) intraperitoneally.

2. Inject animals intravenously with reporter probe [18F]-FHBG at 5 uCi/gm of body weight.

3. Animals are imaged using the microPET system and images processed using ASIPro software.

4. Acquired images between 50 and 75 min are reconstructed by filtered back projection and reoriented into short, vertical, and horizontal axis slices.

5. From regions of interest (ROI) on the anterolateral wall (short axis cut), derived counts/pixel per min are converted to counts/ml per min using a calibration constant obtained from scanning a cylindric phantom.

6. The ROI counts/ml per min are converted to counts/g per min (assuming a tissue density of 1 g/ml) and divided by the injected dose to obtain an image ROI-derived [^{18}F]-FHBG percentage injected dose per gram of heart (% ID/g) as described by Wu et al. and Cao et al. (29, 30).

Reference

1. Cao F, *et al.* (2008) Transcriptional and functional profiling of human embryonic stem cell-derived cardiomyocytes. *PLoS One* 3(10): e3474.

2. Schulz TC, *et al.* (2003) Directed neuronal differentiation of human embryonic stem cells. *BMC Neurosci* 4:27.

3. Jiang J, *et al.* (2007) Generation of insulin-producing islet-like clusters from human embryonic stem cells. *Stem Cells* 25(8): 1940–1953.

4. Swijnenburg RJ, van der Bogt KE, Sheikh AY, Cao F, & Wu JC (2007) Clinical hurdles for the transplantation of cardiomyocytes derived from human embryonic stem cells: role of molecular imaging. *Curr Opin Biotechnol* 18(1):38–45.

5. Kraitchman DL, *et al.* (2003) In vivo magnetic resonance imaging of mesenchymal stem cells in myocardial infarction. *Circulation* 107(18):2290–2293.

6. Li Z, *et al.* (2008) Comparison of reporter gene and iron particle labeling for tracking fate of human embryonic stem cells and differentiated endothelial cells in living subjects. *Stem Cells* 26(4):864–873.

7. Hofmann M, *et al.* (2005) Monitoring of bone marrow cell homing into the infarcted human myocardium. *Circulation* 111(17):2198–2202.

8. Zhang SJ & Wu JC (2007) Comparison of imaging techniques for tracking cardiac stem cell therapy. *J Nucl Med* 48(12):1916–1919.

9. Wu JC, *et al.* (2006) Proteomic analysis of reporter genes for molecular imaging of transplanted embryonic stem cells. *Proteomics* 6(23):6234–6249.

10. Wu JC, *et al.* (2006) Transcriptional profiling of reporter genes used for molecular imaging of embryonic stem cell transplantation. *Physiol Genomics* 25(1):29–38.

11. Zhou R, Acton PD, & Ferrari VA (2006) Imaging stem cells implanted in infarcted myocardium. *J Am Coll Cardiol* 48(10): 2094–2106.

12. Evans MJ & Kaufman MH (1981) Establishment in culture of pluripotential cells from mouse embryos. *Nature* 292(5819): 154–156.

13. Martin GR (1981) Isolation of a pluripotent cell line from early mouse embryos cultured in medium conditioned by teratocarcinoma stem cells. *Proc Natl Acad Sci U S A* 78(12): 7634–7638.

14. Thomson JA, *et al.* (1998) Embryonic stem cell lines derived from human blastocysts. *Science* 282(5391):1145–1147.

15. Li Z, *et al.* (2007) Differentiation, survival, and function of embryonic stem cell derived endothelial cells for ischemic heart disease. *Circulation* 116(11 Suppl):I46–I54.

16. Min JY, *et al.* (2003) Long-term improvement of cardiac function in rats after infarction by transplantation of embryonic stem cells. *J Thorac Cardiovasc Surg* 125(2):361–369.

17. Robinson AJ, *et al.* (2005) Survival and engraftment of mouse embryonic stem cell-derived implants in the guinea pig brain. *Neurosci Res* 53(2):161–168.

18. Duan Y, *et al.* (2007) Differentiation and enrichment of hepatocyte-like cells from human embryonic stem cells in vitro and in vivo. *Stem Cells* 25(12):3058–3068.

19. van der Bogt KE, *et al.* (2008) Comparison of different adult stem cell types for treatment of myocardial ischemia. *Circulation* 118(14 Suppl):S121–S129.

20. Reinecke H & Murry CE (2002) Taking the death toll after cardiomyocyte grafting: a reminder of the importance of quantitative biology. *J Mol Cell Cardiol* 34(3): 251–253.

21. Li L, *et al.* (2004) Human embryonic stem cells possess immune-privileged properties. *Stem Cells* 22(4):448–456.

22. Drukker M, *et al.* (2006) Human embryonic stem cells and their differentiated derivatives are less susceptible to immune rejection than adult cells. *Stem Cells* 24(2):221–229.

23. Swijnenburg RJ, *et al.* (2008) In vivo imaging of embryonic stem cells reveals patterns of survival and rejection following transplantation. *Stem Cells Dev* 17:1023–1029.

24. Swijnenburg RJ, *et al.* (2008) Immunosuppressive therapy mitigates immunological rejection of human embryonic stem cell xenografts. *Proc Natl Acad Sci U S A* 105(35): 12991–12996.

25. Cao F, *et al.* (2007) Spatial and temporal kinetics of teratoma formation from murine embryonic stem cell transplantation. *Stem Cells Dev* 16:1–9.

26. Cao F, *et al.* (2007) Molecular imaging of embryonic stem cell misbehavior and suicide gene ablation. *Cloning Stem Cells* 9(1): 107–117.

27. Ray P, Tsien R, & Gambhir SS (2007) Construction and validation of improved triple fusion reporter gene vectors for molecular imaging of living subjects. *Cancer Res* 67(7): 3085–3093.

28. Loening AM & Gambhir SS (2003) AMIDE: a free software tool for multimodality medical image analysis. *Mol Imaging* 2(3):131–137.

29. Wu JC, Inubushi M, Sundaresan G, Schelbert HR, & Gambhir SS (2002) Positron emission tomography imaging of cardiac reporter gene expression in living rats. *Circulation* 106(2): 180–183.

30. Cao F, *et al.* (2006) In vivo visualization of embryonic stem cell survival, proliferation, and migration after cardiac delivery. *Circulation* 113(7):1005–1014.

Chapter 13

Genetic Fate-Mapping for Studying Adult Cardiomyocyte Replenishment After Myocardial Injury

Sunny S.-K. Chan, Ying-Zhang Shueh, Nirma Bustamante, Shih-Jung Tsai, Hua-Lin Wu, Jyh-Hong Chen, and Patrick C.H. Hsieh

Abstract

Mounting evidence suggests the regenerative potential of the mammalian heart. Nevertheless, the contribution of endogenous stem or precursor cells to adult cardiac regeneration upon myocardial injuries remains unclear. We hereby describe a genetic fate-mapping approach to study adult cardiomyocyte replenishment after myocardial injury. Using double transgenic MerCreMer–ZEG mice, the fate of adult cardiomyocytes can be tracked by the expression of green fluorescence protein (GFP) specifically induced in cardiomyocytes. Upon experimental myocardial infarction, a reduction in GFP expression in the myocardium is observed, indicating the refreshment of cardiomyocytes by endogenous stem or precursor cells.

Key words: Green fluorescence protein (GFP), Genetic fate-mapping

1. Introduction

According to World Health Organization, in 2005, 17.5 million people died of cardiovascular diseases, contributing to 30% of death worldwide (1). Heart failure is chiefly characterized by a 232-fold increase in irreversible myocyte necrosis and apoptosis, which eventually leads to the progression of cardiac dysfunction (2). Although it was first thought that the adult mammalian heart is a postmitotic organ with no regenerative capacity, recent discoveries have established the existence of adult cardiac stem cells. Beltrami et al. recently discovered the existence of stem cells in the heart. They took cells found in the interstitial space between well-differentiated cardiomyocytes which were positive for c-kit, a marker common in stem cells, and analyzed them. Many of these cells expressed transcription factors associated with early

Randall J. Lee (ed.), *Stem Cells for Myocardial Regeneration: Methods and Protocols*, Methods in Molecular Biology, vol. 660, DOI 10.1007/978-1-60761-705-1_13, © Springer Science+Business Media, LLC 2010

cardiac development. While the origin of these cells is unknown, they are self-renewing, clonogenic, multipotent, and give rise to cardiomyocytes, smooth muscle cells, and endothelial cells (3). Laugwitz et al. also discovered Isl1+ cells that may participate in regenerative pathways in newborn heart tissues, giving rise to over one-thirds of the heart (4). The existence of these cells might ignite some enthusiasm, but the extent of their contribution to cardiomyocyte renewal is still controversial. Some fundamental questions remain unanswered. Are cardiomyocytes constantly replaced by endogenous stem or precursor cells? Does injury lead to replacement with new cardiomyocytes from a stem cell pool (5)? Although techniques such as dye-labeling, DNA electroporation, and cell grafting could all be used for cell fate-mapping, disadvantages overshadow their benefits. Dilution of cell marker with every cell division and the fact that some of these techniques require a specific time frame in which cells can be analyzed are some of the reasons why genetic fate-mapping is a far better approach to analyze the fate of stem cells, especially in dealing with the adult mammalian cardiomyocytes (6–8). The following protocol describes, in details, our genetic fate-mapping approach to use a transgenic MerCreMer–ZEG mouse model to demonstrate stem cell refreshment of adult mammalian cardiomyocytes after myocardial infarction.

2. Materials

2.1. Mouse Breeding

1. B6129-Tg(*Myh6-cre/Esr1*)1Jmk/J mice (Jackson Laboratory, stock number 005657).
2. B6.Cg-Tg(*CAG-Bgeo/GFP*)21Lbe/J mice (Jackson Laboratory, stock number 004178).
3. Animal housing equipment: sterilized water, standard mouse chow, and a specific pathogen-free mouse housing facility.

2.2. Genotyping

1. Genotyping PCR kit (Sigma REDExtract-N-Amp Tissue PCR Kit, XNATS).
2. Primers: MerCreMer forward primer: 5′-GTCTGACTAGGT GTCCTTCT-3′; MerCreMer reverse primer: 5′-CGTCCT CCTGCTGGTATAG-3′; ZEG forward primer: 5′-AAGTTCA TCTGCACCACCG-3′; ZEG reverse primer: 5′-TCCTTG AAGAAGATGGTGCG-3′; Interleukin-2 forward primer: 5′-CTAGGCCACAGAATTGAAAGATCT-3′; Interleukin-2 reverse primer: 5′-GTAGGTGGAAATTCTAGCATCATCC-3′. Interleukin-2 acts as an internal control. Dissolve all primers in water to 20 µM before use.

3. Tris-Borate-EDTA (TBE) buffer (10×): 108 g Tris base, 55 g boric acid, and 9.3 g Na_4EDTA in 1 l water. The pH is 8.3 and requires no adjustment. Dilute to 0.5× in water before use.

4. Agarose, 2% (Cambrex Bio Science): 0.64 g of agarose in 32 ml of 0.5% TBE buffer. Microwave 2–3 min for complete dissolution.

5. Ethidium bromide (Sigma): 0.5 µg/ml in water before use.

2.3. Tamoxifen Injection

1. Tamoxifen (Sigma): 1 g of tamoxifen in 10 ml absolute ethanol. Add 90 ml sunflower seed oil and mix well for complete dissolution (final concentration = 1 mg/100 µl). Store at –20°C.

2.4. Experimental Myocardial Ischemia

1. Pentobarbital sodium (Sigma): 40 mg/ml in water.

2. Electric shaver.

3. Polyethylene tubing (PE-180) for tracheal intubation.

4. Artificial ventilator (Harvard Apparatus).

5. Surgical tools: surgical scissors, tissue forceps, retractors.

6. Suture: Prolene 6–0 polypropylene suture, Ethibond Excel 5–0 polyester suture, Silk 5–0 silk suture (all from Ethicon).

7. Heating pad.

2.5. Bromodeoxyuridine Perfusion

1. Osmotic minipump (Alzet).

2. Bromodeoxyuridine (BrdU) (Sigma).

2.6. Heart Harvesting, Fixation, Dehydration, and Embedding

1. Phosphate buffered saline: 8 g NaCl, 0.2 g KCl, 1.44 g Na_2PO_4, and 0.24 g KH_2PO_4 in 1 l water. Adjust to pH 7.4 with HCl if necessary.

2. Fixation reagents: ethanol, xylene, paraffin.

3. Embedding mold cassettes.

2.7. Immunohistochemistry and Immunofluorescence Microscopy

1. Sodium citrate solution (10 mM): 2.94 g sodium citrate in 1 l water, pH 6.0.

2. H_2O_2 (3%): 10 ml 30% H_2O_2 in 90 ml water. Freshly prepare before use.

3. Wash buffer: 0.1% Tween-20 in phosphate buffered saline (pH 7.4).

4. Blocking Solution: 5% fetal bovine serum and 5% goat serum in wash buffer.

5. Primary antibodies: Anti-GFP (1:200, Abcam), anti-β-galactosidase (1:200, Invitrogen), anti-tropomyosin (1:100, Developmental Studies Hybridoma Bank), anti-cardiac troponin-T (1:100, Developmental Studies Hybridoma Bank),

anti-α-myosin heavy chain (1:100, Abcam), anti-α-smooth muscle actin (1:100, Sigma), anti-BrdU (1:50, Roche).

6. Secondary antibody: Alexa Fluor (Invitrogen).

7. 4′,6-Diamidino-2-phenylindole (DAPI): 1 µg/ml in phosphate buffered saline.

2.8. Counting GFP⁺ and GFP⁻ Cardiomyocytes

1. ImageProbe software (ImagiWorks).

3. Methods

3.1. Generation of Double-Transgenic MerCreMer–ZEG Mice

1. Generate double transgenic MerCreMer–ZEG mice by crossbreeding transgenic B6129-Tg(*Myh6-cre/Esr1*)1Jmk/J (hereafter referred to as MerCreMer) mice and B6.Cg-Tg(*CAG-Bgeo/GFP*)21Lbe/J (hereafter referred to as ZEG) mice (Fig. 1). MerCreMer mice contain a tamoxifen-inducible Cre recombinase fusion protein driven by the cardiomyocyte-

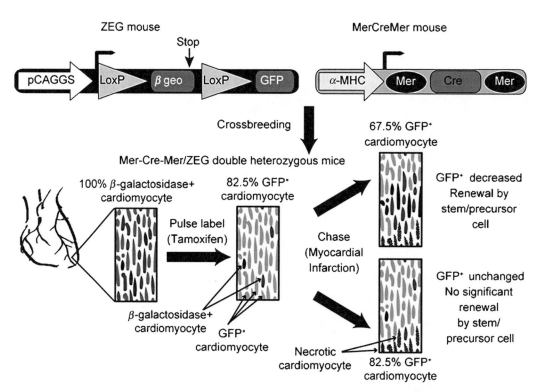

Fig. 1. Generation of double transgenic MerCreMer–ZEG mice. Double transgenic MerCreMer–ZEG mice are generated by crossbreeding MerCreMer and ZEG mice. Upon tamoxifen induction (pulse label), the Cre recombinase fusion protein driven by α-myosin heavy chain (MHC) promoter provokes the replacement of β-galactosidase (β geo) expression with GFP in cardiomyocytes. A decrease in the percentage of GFP⁺ cardiomyocytes upon myocardial infarction (chase) suggests a possible cardiomyocyte replenishment by stem or progenitor cells.

specific α-myosin heavy chain promoter. In ZEG mice, green fluorescence protein (GFP) replaces constitutive β-galactosidase expression after the removal of a *loxP*-flanked stop sequence.

2. Start mating when mice reach at least 8 weeks of age, with one male housed with two females (one MerCreMer male with two ZEG females or one ZEG male with two MerCreMer females).

3. Determine pregnancy by the presence of a copulatory plug (a cream-colored plug of solidified ejaculate) in the female vagina. Remove the male after pregnancy is confirmed. If pregnancy does not occur within 14 days of pairing, reshuffle the mating pair.

4. A litter of 6–12 pups is expected to be delivered after 21 days. Neonatal mice are fragile and extra care is needed when handling (see Note 1).

5. Determine the genotype and wean mice at 21 days of age.

6. During the whole course of housing, maintain a room temperature of ~22°C, relative humidity of 40–60%, and a 12-h/12-h light–dark cycle. Also, change bedding once a week and allow the animals free access to sterilized water and standard mouse chow.

3.2. Genotyping

3.2.1. Tissue Preparation

1. Determine mice genotype by PCR on ear tissues using a commercial kit (Sigma REDExtract-N-Amp Tissue PCR Kit, XNATS) (see Note 2).

2. Obtain ear tissue with a ear hole plunger. Rinse the plunger and forceps in 70% ethanol prior to use and between different samples to prevent cross-contamination.

3. Prepare a mixture of 100 μl Extraction Solution and 25 μl Tissue Preparation Solution in a microcentrifuge tube by pipetting up and down several times. Immerse the ear tissue (1 mm in diameter) into the solution and mix thoroughly by vortexing.

4. Incubate at room temperature for 10 min and then at 95°C for 3 min. It is normal if tissues are not completely digested at the end of incubation.

5. Add 100 μl of Neutralization Solution B to stop digestion and mix well.

6. Store the neutralized tissue extract at 4°C or use immediately for PCR. Extracts are stable at 4°C for at least 6 months.

3.2.2. PCR Amplification

1. Prepare the following in a PCR microcentrifuge and mix gently: 5.2 μl PCR grade water, 10 μl REDExtract-N-Amp PCR Reaction Mix, 0.4 μl forward primer (20 μM), 0.4 μl reverse primer (20 μM), and 4 μl tissue extract.

2. Perform PCR amplification as follows:

Step	Temperature(°C)	Time (min)	Cycle
Initial denaturation	94	3	1
Denaturation	94	0.5	30
Annealing	53	0.5	30
Extension	72	0.5	30
Final extension	72	10	1
Hold	4	Indefinitely	

3. PCR products can be used immediately for electrophoresis or stored at 4°C.

3.2.3. Electrophoresis

1. Prepare a 2% (w/w) agarose solution by adding 0.64 g of agarose in 32 ml of 0.5× TBE buffer. Microwave 2–3 min for complete dissolution.

2. Pour the hot agarose solution into a gel tray and allow it to settle for at least an hour.

3. Load 8 μl of PCR products into the agarose gel and run at 100 V for 25 min.

4. Bath the gel into ethidium bromide (0.5 μg/ml) for 5 min for DNA staining.

5. Wash the gel with water and examine the DNA bands under ultra violet light at 254 nm wavelength.

6. Only mice with a genotype of MerCreMer–ZEG will be used for subsequent study.

7. Figure 2 shows a typical genotyping result.

3.3. Tamoxifen Injection

1. Dissolve 1 g of tamoxifen (Sigma) in 10 ml of absolute ethanol.

2. Add 90 ml of sunflower seed oil and mix well for complete dissolution (final concentration = 1 mg/100 μl). Store at –20°C.

3. Inject 100 μl of tamoxifen solution intraperitoneally into 8-week-old MerCreMer–ZEG mice daily for 14 consecutive days. More than 95% MerCreMer–ZEG mice are expected to survive following a full course of tamoxifen injection (see Note 3).

3.4. Experimental Myocardial Ischemia

1. Seven days after the last tamoxifen injection, MerCreMer–ZEG mice are subjected to experimental myocardial infarction.

2. Anesthetize the animal with intraperitoneal administration of pentobarbital sodium (40 mg/kg). Adequate anesthesia is defined as a loss of righting and withdrawal reflexes (see Note 4).

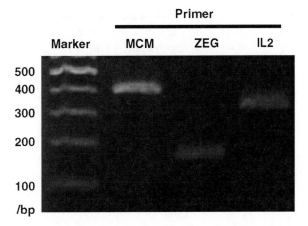

Fig. 2. A typical genotyping result of MerCreMer–ZEG mice. Bands at 410 bp (*Lane 2*), 173 bp (*Lane 3*) and 324 bp (*Lane 4*) correspond to the presence of MerCreMer (MCM), ZEG and interleukin-2 (IL2) genes in a MerCreMer–ZEG mouse. Interleukin-2 acts as an internal control.

3. Remove the fur on the chest and the surrounding area using an electric shaver.

4. Place the anesthetized animal in a supine position and intubate the trachea with a 5-cm long polyethylene tubing (PE-180) connected to an artificial ventilator at a respiratory rate of ~120 breaths/min and a stroke volume of ~0.5 ml/breath.

5. Make a 2-cm long incision 0.5 cm to the left of the chest midline along the sternum. Blunt dissect the subcutaneous tissues to expose the rib cage.

6. Make a 1.5-cm incision at the fourth intercostal space. Place a retractor to spread the ribs apart for a better view of the heart.

7. Locate the left anterior descending coronary artery which lies between the superficial landmarks of the left atrial appendage and the right outflow tract. Place an Ethicon Prolene 6–0 polypropylene suture attached to a reverse cutting needle around the left anterior descending coronary artery at ~2 mm distal to the left atrial appendage.

8. Tie the ends of the suture a few times to secure complete ligation of the left anterior descending coronary artery. Successful myocardial ischemia is determined by discoloration and dyskinesia of the myocardium distal to the ligation site.

9. Close the rib cage with an Ethicon Ethibond Excel 5–0 polyester suture and suture the skin with an Ethicon Silk 5–0 silk suture.

10. Disconnect the ventilator and check for spontaneous breathing. Reconnect to the ventilator if spontaneous breathing does not occur within a few seconds.

11. Remove the tracheal intubation tubing and place the animal onto a heating pad.

12. Put the animal back to the cage once consciousness is gained (see Note 5).

3.5. Bromodeoxy-uridine Perfusion

1. At the time of myocardial ischemia surgery, implant an Alzet osmotic minipump subcutaneously.

2. Remove the fur on the middle of the back using an electric shaver.

3. Make a 1-cm long incision along the midline in the back. Blunt dissect the subcutaneous tissues to create a pocket.

4. Insert the pump, which deliver BrdU (Sigma) at 1 μg/h/g body weight for 7 days, into the pocket (see Note 6).

5. Suture the skin with an Ethicon Silk 5-0 silk suture.

3.6. Heart Harvesting, Fixation, Dehydration and Embedding

1. Three months after surgery, euthanize mice with CO_2 asphyxiation.

2. Harvest the heart, wash with phosphate buffered saline, and fix with 4% formaldehyde for 16 h in a 20-ml snap-cap glass vial at 4°C.

3. After fixation, in a chemical hood, remove the fixative, and immerse the fixed heart into 70% ethanol for 24 h.

4. Immerse in 95% ethanol for 30 min twice, following by 100% ethanol for 30 min twice, and xylene for 15 min four times. Xylene is very toxic and all steps should be performed in a chemical hood.

5. Remove xylene and add in molten paraffin (previous melted in 58°C oven). Incubate for 45 min at 58°C. Repeat thrice.

6. Transfer the sample into embedding metal mold cassette filled with molten paraffin.

7. Place an embedding ring on the mold and fill with paraffin wax.

8. Leave the mold block at room temperature for hardening.

9. Remove the mold cassette and store the embedded block at a dry place.

3.7. Immunohisto-chemistry and Immunofluorescence Microscopy

1. Deparaffinize fixed paraffin-embedded heart sections by immersing into xylene for 5 min twice, 100% ethanol for 3 min twice, 95% ethanol for 3 min twice, and 70% ethanol for 3 min twice.

2. Rehydrate sections by washing in water for 3 min twice.

3. Immerse sections in boiling 10 mM sodium citrate buffer (pH 6.0) for 10 min, and subsequently cool the sections at room temperature for 30 min. Wash with water for 5 min thrice.

4. Immerse sections in 3% H_2O_2 for 10 min. Wash with water for 3 min twice, following by wash buffer (0.1% Tween-20 in phosphate buffered saline, pH 7.4) for 3 min.

5. Immerse sections in blocking buffer (5% fetal bovine serum and 5% goat serum in wash buffer) for 1 h at room temperature.

6. Remove blocking buffer, add anti-GFP primary antibody (1:1,000, MBL), and incubate overnight at 4°C. Wash with wash buffer for 3 min twice.

7. Add Alexa Fluor secondary antibody (1:200, Invitrogen) and incubate for 30 min at room temperature. Wash with wash buffer for 3 min twice.

8. Repeat steps 6 and 7 for costaining with other primary antibodies such as anti-β-galactosidase (1:500, Invitrogen), anti-tropomyosin (1:100, Developmental Studies Hybridoma Bank), anti-cardiac troponin-T (1:100, Developmental Studies Hybridoma Bank), anti-α-myosin heavy chain (1:100, Abcam), or anti-α -smooth muscle actin (1:100, Sigma) and another Alexa Fluor secondary antibody (Invitrogen) for counterstaining.

9. For BrdU costaining, an addition step of DNA denaturation is required. Denature DNA with 2N HCl for 1 h at 37°C, follow by acid neutralization with 0.1 M borate buffer for 10 min twice, and wash with phosphate-buffered saline for 5 min twice. Repeat steps 6 and 7 with adding anti-BrdU primary antibody (1:50, Roche) and incubating for 1 h at room temperature.

10. Counterstain by adding DAPI (4′,6-diamidino-2-phenylindole, 1 µg/ml in phosphate buffered saline) for 1 min at room temperature. Wash with water for 3 min twice.

11. Dehydrate by immersing into 70% ethanol for 30 s twice, following by 95% ethanol 30 s twice, 100% ethanol 30 s twice, and xylene 1 min twice.

12. Mount coverslip and count GFP+ and GFP- cardiomyocytes under fluorescence microscope.

3.8. Counting GFP+ and GFP- Cardiomyocytes

1. Take 3 sections from each heart and locate two infarction border zones from each section.

2. Examine each border zone under fluorescence microscope, and use the ImageProbe software for color separation to increase contrast.

Anti-GFP   **Anti-tropomyosin**

DAPI   **Merged**

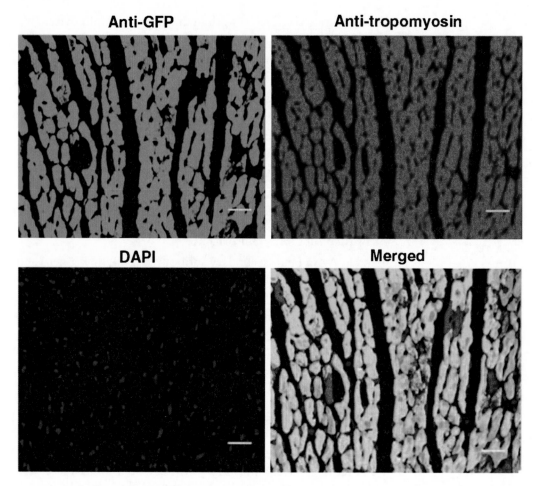

Fig. 3. A typical GFP-tropomyosin-DAPI triple staining in MerCreMer–ZEG mouse myocardium. Note that most cardiomyocytes, as labeled by anti-tropomyosin, express GFP. Scale bars, 20 μm.

3. Count the number of GFP$^+$ and GFP$^-$ cells with visible sarcomere (cardiomyocytes) by a "blinded" person without a priori knowledge of treatments.

4. Count at least 300 cardiomyocytes for each heart.

5. A typical stain is shown in Fig. 3.

4. Notes

1. Young mothers (8–10 weeks of age) sometimes fail to take enough care of their first-borns. Housing with a fostering mother may be necessary. In any case, mothers usually provide adequate care to their pups starting from the second-born.

2. We have found this kit very time-efficient and excellent to use, although other genotyping methods can also be used.

3. 4-OH-Tamoxifen can be used instead of tamoxifen. In fact, 4-OH-tamoxifen produces almost identical results to tamoxifen at half the dosage (i.e., injection of 100 µl at 0.5 mg/100 µl/day for 14 days), but 4-OH-tamoxifen is much more expensive.

4. It is important to avoid deep anesthesia which will jeopardize the animal recovery after subsequent highly invasive myocardial ischemia surgery.

5. Animals usually regain consciousness at a faster rate when put inside an oxygen chamber.

6. Alternatively, BrdU can be administered intraperitoneally at 25 µg/g body weight daily for 7 days.

Acknowledgments

This work is supported by grants from the National Science Council (NSC 97IR082), the National Heath Research Institutes (NHRI EX97-9722SI) and the NCKU Landmark and Integrative Research Projects (NCKU 971006).

References

1. "Statistical Fact Sheet – Populations." American Heart Association. 2008. June 16, 2008. http://americanheart.org/downloadable/heart/1201543457735FS06INT08.pdf.

2. Olivetti, G., Abbi, R., Quaini, F., Kajstura, J., Cheng, W., Nitahara, J. A., Quaini, E., Di Loreto, C., Beltrami, C. A., Krajewski, S., Reed, J. C., and Anversa, P. (1997) Apoptosis in the failing human heart. *N Engl J Med* **336**, 1131–1141.

3. Beltrami, A. P., Barlucchi, L., Torella, D., Baker, M., Limana, F., Chimenti, S., Kasahara, H., Rota, M., Musso, E., Urbanek, K., Leri, A., Kajstura, J., Nadal-Ginard, B., and Anversa, P. (2003) Adult cardiac stem cells are multipotent and support myocardial regeneration. *Cell* **114**, 763–776.

4. Laugwitz, K. L., Moretti, A., Lam, J., Gruber, P., Chen, Y., Woodard, S., Lin, L. Z., Cai, C. L., Lu, M. M., Reth, M., Platoshyn, O., Yuan, J. X., Evans, S., and Chien, K. R. (2005) Postnatal isl1+ cardioblasts enter fully differentiated cardiomyocyte lineages. *Nature* **433**, 647–653.

5. Hsieh, P. C. H., Segers, V. F. M., Davis, M. E., MacGillivray, C., Gannon, J., Molkentin, J. D., Robbins, J., and Lee, R. T. (2007) Evidence from genetic fate-mapping study that stem cells refresh adult mammalian cardiomyocytes after injury. *Nat Med* **13**, 970–974.

6. Bildsoe, H., Frankling, V., and Tam, P. P. L. (2007) Fate-mapping technique: using carbocyanine dyes for vital labeling of cells in gastrula-stage mouse embryos cultured in vitro. *CSH Protocols* doi:10.1101/pbd.prot4915.

7. Bildsoe, H., Frankling, V., and Tam, P. P. L. (2007) Fate-mapping technique: grafting fluorescent cells into gastrula-stage mouse embryos at 7–7.5 days post-coitum. *CSH Protocols* doi:10.1101/pbd.prot4892.

8. Khoo, P. L., Franklin, V., and Tam, P. P. L. (2007) Fate-mapping technique: targeted whole-embryo electroporation of DNA constructs into the germ layers of mouse embryos 7–7.5 days post-coitum. *CSH Protocols* doi:10.1101/pbd.prot4893.

Part V

Electrophysiology

Chapter 14

In Vitro Electrophysiological Mapping of Stem Cells

Seth Weinberg, Elizabeth A. Lipke, and Leslie Tung

Abstract

The use of stem cells for cardiac regeneration is a revolutionary, emerging research area. For proper function as replacement tissue, stem cell-derived cardiomyocytes (SC-CMs) must electrically couple with the host cardiac tissue. Electrophysiological mapping techniques, including microelectrode array (MEA) and optical mapping, have been developed to study cardiomyocytes and cardiac cell monolayers, and these can be applied to study stem cells and SC-CMs. MEA recordings take extracellular measurements at numerous points across a small area of cell cultures and are used to assess electrical propagation during cell culture. Optical mapping uses fluorescent dyes to monitor electrophysiological changes in cells, most commonly transmembrane potential and intracellular calcium, and can be easily scaled to areas of different sizes. The materials and methods for MEA and optical mapping are presented here, together with detailed notes on their use, design, and fabrication. We also provide examples of voltage and calcium maps of mouse embryonic stem cell-derived cardiomyocytes (mESC-CMs), obtained in our laboratory using optical mapping techniques.

Key words: Electrophysiology, Microelectrode array, Optical mapping, Embryonic stem cell, Cardiomyocyte

1. Introduction

Cardiac regeneration involves the repair of damaged or diseased cardiac tissue. If stem cells and stem cell-derived cardiomyocytes (SC-CMs) are to be successful as replacement cells and tissue, they must possess key electrical properties that permit them to integrate with the host tissue either individually or as tissue constructs. It will be essential to assess the electrophysiological properties of these cells. Specialized mapping techniques and tools have been developed over the years to study cardiomyocytes and cardiac cell monolayers and can be applied to study SC-CMs. Knowledge generated using these tools will be key to creating tissue constructs that are electrophysiologically compatible and

Randall J. Lee (ed.), *Stem Cells for Myocardial Regeneration: Methods and Protocols*, Methods in Molecular Biology, vol. 660,
DOI 10.1007/978-1-60761-705-1_14, © Springer Science+Business Media, LLC 2010

functionally coupled with recipient myocardium. Pertinent questions include: do SC-CMs electrically couple with each other and with native cardiomyocytes, can the electrophysiological properties of SC-CMs and SC-CM tissues recapitulate those of fully developed cardiac tissue, do physiologically relevant-sized sheets of SC-CMs support coordinated wave propagation, and does inappropriate coupling and/or autorhythmic activity give rise to cardiac arrhythmias? This chapter will address the experimental techniques and tools that can be used to answer these questions, including the use of microelectrode (sometimes called multielectrode) arrays (MEAs) and optical mapping systems, and will introduce some of the unique challenges faced when studying SC-CM electrophysiology.

MEAs are multiple electrode systems that are used to make extracellular recordings of electrical activation times at numerous sites in small areas (0.01–0.25 cm^2) of freshly sliced tissue sections, organotypic cultures, and cultured cells, including SC-CMs and primary cardiomyocytes (see Note 1). A preparation grown on a MEA can be recorded at multiple times during long-term culture, not just at one time point. The usefulness of MEAs in studying embryonic SC-CMs (1) and cardiomyocytes (2) has been recently reviewed. Numerous studies have used MEA systems to monitor electrical activity in SC-CMs (3–6) and electrical coupling of SC-CMs with cardiomyocytes isolated from cardiac tissue (7, 8). In addition, MEA recordings of human embryonic SC-CMs may eventually provide a high-throughput assay to screen pharmaceutical agents to improve drug efficacy, specificity, and safety (9, 10).

Optical mapping is a technique for measuring the spread of electrical activation and repolarization among excitable cells, such as cardiac, neuronal, or stem cells, and is a valuable tool for measuring changes under numerous experimental conditions, such as drug superfusion, cocultures with other cell types, or electrical stimulation. Mapping has been performed at the microscopic (single or few cells (11)), macroscopic (1D strands (12, 13) and 2D monolayers (14)), and whole heart levels (15–18). Most mapping studies of cardiac cells have used fluorescent dyes owing to the higher fractional changes in signal compared with other modalities such as absorption (19). The two types of dyes that have been most widely used are voltage-sensitive and calcium-sensitive dyes. Transmembrane voltage governs cellular excitation and recovery. Calcium ions carry ionic current through membrane ion channels and transporters and play a key signaling role for excitation-contraction coupling and many other cellular processes. The bidirectional coupling of calcium and transmembrane voltage is now under active investigation because of its potential importance in causing cardiac arrhythmias (20). There also exist functional fluorescent probes which are sensitive to other intracellular signals, such as sodium, magnesium, potassium, pH, cyclic nucleotides, and nitric oxide (Invitrogen, Carlsbad, CA). Mapping of wavefront propagation has also been achieved by dye-free, optical phase imaging that is sensitive to local cell contractile motion (21).

The key optical components for fluorescence mapping are an excitation light source, an emission filter, and a photodetector. In an optical mapping experiment, the cells are stained with a fluorescent dye and then illuminated by the excitation light source. The dye emits fluorescent light at a wavelength longer than that of the excitation light, which passes through an emission filter and is converted by the photodetector into an electrical signal. Optimization of the optical configuration involves judicious choices of the mode of illumination, fluorescent dye(s) used, excitation light source (see Note 2), excitation filter (see Note 3), emission filter, and photodetector system (see Note 4) – in relation to the application requirements.

There are two modes of illumination used for optical mapping: transillumination, meaning that the excitation light and photodetector are on opposite sides of the coverslip containing the cell preparation, and epiillumination, where the excitation light and photodetector are on the same side of the coverslip. A simple lens-less configuration called contact fluorescent imaging (CFI) (14, 22) was developed in our lab (Fig. 1a). In this configuration, which utilizes transillumination, an excitation light source is placed directly above a coverslip on which cells are grown, and one end of an optical fiber bundle is placed in contact with a thin emission filter lying directly beneath the coverslip. The other end couples to a bank of photodiodes (see Note 5). Other mapping configurations use one or more lenses. In a single lens configuration, the optical fiber bundle is replaced with a lens that focuses the light onto the photodetector. In a tandem-lens

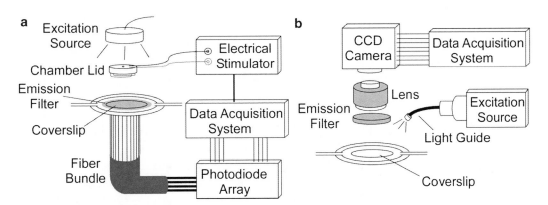

Fig. 1. Schematic of optical mapping configurations. (**a**) Contact fluorescence imaging (CFI). (**b**) Epiillumination with a CCD camera and a single lens. The electrical stimulator and chamber lid (shown in **a**) are also present in the CCD camera configuration but are not shown.

(TL) configuration, two lenses are used to improve the effective numerical aperture (see Note 6). In a configuration utilizing epiillumination, the photodetector and emission filter are both placed on the same side (either above or below) of the cells being mapped. The excitation light source is positioned either at a slight angle from the photodetector to allow direct illumination of the coverslip (Fig. 1b) or at a 90° angle into a reflective dichroic mirror (angled at 45°) which redirects the excitation light onto the cells and also passes only the emitted fluorescent light back to the photodetector. Descriptions, advantages, and other practical design constraints of different optical mapping configurations have been reviewed (23, 24).

For mapping on a microscopic scale (<0.5 mm), a conventional epifluorescence microscope can be used to map small clusters of stem cells or SC-CMs (see Note 7) (25–27). The synchrony (or its absence) of the voltage or calcium signals from neighboring cells reveals either the presence (or absence) of cell–cell coupling. Two-dimensional confocal laser scanning fluorescence imaging has also been used to measure calcium transients from small clusters of cells (28, 29). Because of the long scanning time (on the order of several seconds), confocal imaging is typically used to look at slow changes in calcium signals.

Because the fluorescent signal is much weaker from voltage-sensitive dyes than from calcium-sensitive dyes, it is better with cell cultures to measure the former with photodiode arrays (PDAs) than with a charge-coupled device (CCD) camera, given their higher sensitivity (see Note 4). In our lab, CFI is used for voltage mapping (Fig. 1a) and a CCD camera for calcium mapping (Fig. 1b) of large-sized (~2 cm diameter) cell monolayers, and the materials and methods are described for both systems. Voltage and calcium maps of mouse embryonic stem cell-derived cardiomyocyte (mESC-CM) monolayers are also presented.

2. Materials

2.1. Microelectrode Array Preparation

2.1.1. Microelectrode Array

1. MEA.
2. Cloning cylinder (Fisher Scientific, Pittsburgh, PA): A cloning cylinder is used as a barrier to reduce the working surface area (see Note 8).
3. Fibronectin (Sigma-Aldrich, St. Louis, MO, F0895): 5 μg in 1 mL of phosphate-buffered saline (PBS) solution, stored at 4°C (see Note 9).

2.1.2. Cell Culture

1. Culture media for mouse embryonic SC-CMs (see Note 10): DMEM (Gibco, Invitrogen, 111965-092, includes glutamine);

10–15% fetal bovine calf serum (Atlanta Biologicals, Lawrenceville, GA) (heat-inactivate the serum prior to use: 56°C for 30 min, then cool prior to adding); GlutaMax-1 (Gibco 35050-061) 1 mL/100 mL; MEM non-essential amino acids (Gibco 11140-050) 1 mL/100 mL; sodium pyruvate (Gibco 11360-070) 1 mL/100 mL; 2-mercapto-ethanol (1,000×) (Gibco 21985-023) 1 µL/mL; gentamicin sulfate (BioWhittaker, Lonza, Basel, Switzerland, 17–518 z) 50 µL/100 mL.

2.1.3. MEA System Components

1. Recording chamber (with means to control temperature of MEA during recordings).

2. Electrode amplifiers.

3. Data acquisition system and software.

4. Inverted microscope to observe cells during data acquisition (see Note 11).

5. Anti-vibration table (see Note 12).

6. Tyrode's solution (see Note 13).

2.2. Voltage Mapping Using Contact Fluorescent Imaging

2.2.1. Mapping System Components

1. Experimental chamber (see Note 14): The custom-built chamber contains tubing connections for solution to flow in and out, electrodes for stimulation, and a thermocouple to monitor temperature. The chamber lid is made of acrylic plastic, and the chamber floor is the emission filter, discussed below.

2. Superfusion system (see Note 15): The superfusion system includes Tygon tubing (Cole-Parmer, Vernon Hills, IL) and an eight roller peristaltic pump (REGLO Analog, Ismatec, Glattbrugg, Switzerland) (see Note 16), that flows solution into the chamber at a constant rate of ~2.6 mL/min.

3. Heating system (see Note 17): Solution is heated using a thermocouple-regulated feedback temperature controller (TC-324B, Warner Instruments), placed in series with the tubing connected to the chamber.

4. Tyrode's solution (see Note 13).

5. Electrodes (see Note 18): The monolayer paced by either line or point electrodes. The chamber lid has two platinum (Surepure Chemetals, Florham Park, NJ) wire line electrodes protruding on opposite sides (Fig. 1a). The top of the chamber lid also has a small hole, which allows insertion of a thin platinum point electrode above the cell monolayer.

6. Electrical stimulator (see Note 19): The output of either an SD9 or S48 stimulator (Grass Technologies, West Warwick, RI) provides the current to stimulate the cells.

1. Voltage-Sensitive Dye (see Note 20): di-4-ANEPPS. Di-4-ANEPPS (MW 480.66 g) is prepared as a 4 mM stock solution: Add 2.6 mL dimethylsulfoxide (DMSO) (Sigma-Aldrich) to a 5 mg bottle of di-4-ANEPPS (shipping bottle). Sonicate the bottle at 60°C for 15 min. Aliquot 2.5 µL of the stock solution into 1 mL centrifuge tubes (see Note 21). Cover the sealed bottle and centrifuge tubes with aluminum foil and store in the refrigerator.

2. Light-emitting diode (LED) excitation light source: The excitation source consists of a custom-built array of 26 high-power green LEDs (WP7104, Kingbright, Taipei, Taiwan), powered at 30 mA and arranged in a pattern of three concentric circles (see Note 22). The LED array and excitation filter are housed in a custom-designed aluminum metal casing that is mounted 2 cm above the chamber, facing straight down toward the chamber during the experiment. The mount is on a swivel arm so that it can be moved in and out of position to allow easy access to the chamber. Mechanical stops assure that the arm is in the same position each time it is over the chamber.

3. Excitation filter: The LED array is arranged behind a 35 mm diameter interference filter (530 ± 25 nm, Chroma Technology, Rockingham, VT).

4. Custom spin-coated red emission filter: Our lab uses a red emission filter, which passes wavelengths longer than ~590 nm (see Note 23). To minimize its thickness (see Note 24), the filter is a custom-made spin-coated glass coverslip (30 mm diameter, No. 1 thickness, 130–160 µm). To fabricate the filter, place a coverslip on a spin-coater (Laurell Technologies, North Wales, PA). Use a plastic pipette to place several drops of red ink (Avery Denison, Brea, CA) or photoresist (PSCRed, Brewer Science, Rolla, MO) in the middle of the coverslip. Spin at 2,000 rpm for 60 s and in the case of photoresist, then bake the coverslip at 125°C for 60 s. Repeat spinning and baking two more times, such that there are three total layers. Mount the filter as the chamber floor, with the photoresist-side facing downward, away from the bath solution.

5. Optical fiber bundle: The custom-built fiber bundle is composed of 253 plastic optical fibers (1 mm diameter), packed in a 17 mm diameter hexagonal array. The bundle is fabricated by using a jig to squeeze the fibers together in the proper packed configuration after they are coated with optical epoxy. Once the epoxy has hardened, the face of the bundle is optically polished using increasingly fine lapping paper.

6. PDA and instrumentation: Each optical fiber is coupled to an individual photodetector that consists of a photodiode (S6786, Hamamatsu, Japan), a current-to-voltage converter

(OPA124, Burr-Brown, Langhorne, PA), and filtering and amplification stages (see Note 25).

7. Data acquisition system and software (see Note 26): Individual analog signals are sampled at 1 kHz (see Note 27) and digitized with 16-bit data acquisition boards (Sheldon Instruments, San Diego, CA). The digital signals are acquired using custom-written LabVIEW (National Instruments, Austin, TX) software.

2.3. Calcium Mapping Using a Single Lens and a CCD Camera

2.3.1. Mapping System Components (Same as Described in Subheading 2.2.1 for CFI)

2.3.1.1. Optical Components

1. Calcium-Sensitive Dye (see Note 28): Rhod-2-AM. Rhod-2-AM (MW 1123.96 g) is prepared as a 1 mM stock solution; Pluronic is prepared as a 4% w/v stock solution: The dye typically is packaged in 50 μg vials. Add 44.5 μL DMSO to the vial. Sonicate the vial at 50°C for 30 min. Pipette out the stock solution into 5 μL aliquots in 1 mL centrifuge tubes or other small containers. Wrap each tube with Parafilm and aluminum foil and store in the freezer. Dissolve 0.04 g Pluronic F-127 in 1 mL DMSO in a 15 mL conical tube. Place the solution in warm water bath for 20 min, until Pluronic is thoroughly dissolved. Store at room temperature.

2. Quartz-tungsten halogen (QTH) lamp excitation light source: The excitation light consists of the output of a 250 W QTH lamp (Oriel, Newport Corporation, Irvine, CA), directed onto the coverslip via a 5 mm diameter liquid light guide (Oriel).

3. Excitation filter: A 45 mm diameter interference filter (535 ± 25 nm, Chroma Technology) (see Note 29) is placed in the QTH lamp housing.

4. CCD camera (see Note 30): Andor (South Windsor, CT) iXon+ 860 EMCCD (128×128 pixels). CCD images are acquired using custom-written LabVIEW software.

5. Lens (see Note 31): Cinegon f/1.4 (Schneider Optics, Germany) lens.

6. Emission filter: Bandpass (580 ± 5 nm, XB101, Omega Optical, Brattleboro, VT) filter (see Note 32). The filter is placed in a lens filter screw-in ring to be mounted in front of the camera lens.

3. Methods

3.1. Microelectrode Array Recordings

3.1.1. Preparation of MEA

1. Sterilize the MEA according to manufacturer's instructions, for example, by autoclaving.

2. Place the sterilized cloning cylinder around the area containing the electrodes.

3. Coat the surface of the MEA with fibronectin (see Note 33).

3.1.2. Seeding and Culture of Cells

1. Seed the cells at the density needed for confluency or seed undissociated cells, for example, as an intact embryoid body, according to the experimental design. The cells need to make tight contact with the electrodes for good recordings.

2. Place a few small drops of sterile water around the outside of the cloning cylinder to maintain humidity (see Note 34).

3. Cover the top of the MEA with a sterile glass lid and place into the incubator (see Note 35).

4. Culture the cells for 24–48 h before initiating recordings. Replace the medium as usual. Use a sterile forceps to remove the cloning cylinder as soon as the cells are firmly adherent (12–24 h) and add medium to prevent the culture from becoming acidic. Aspirate any water droplets outside the cloning cylinder prior to its removal.

3.1.3. MEA Setup and Recording

1. Turn on the MEA system. Allow the recording chamber/amplifier to equilibrate to 37°C.

2. Clean the MEA contact pads with alcohol to remove any residue. Place the MEA into the amplifier, using the microscope to check that the MEA is oriented correctly with respect to the electrode array. Make note of the electrode spacing for postexperiment analysis. Ground the bath.

3. When making recordings from the same MEA on multiple days, filter sterilize the Tyrode's solution prior to use. It is important to minimize the temperature variation. For this reason, keep the Tyrode's solution in a 37°C water bath until use. Immediately before initiation of recording, remove the culture medium from the MEA and replace with sterile Tyrode's solution (see Note 36).

4. Observe the cells through the microscope and note the location of any areas without cells or other inconsistencies. Take a photomicrograph to document the location of the cells, including areas of higher or lower cell density (see Note 37).

5. When the temperature of the MEA has restabilized at 37°C, take recordings based on the desired protocol (see Note 38).

6. Data files can be analyzed using custom MATLAB (MathWorks, Natick, MA) scripts; one resource for data analysis is MEA-Tools, a free downloadable MATLAB toolbox (30) (see Note 39).

3.2. Voltage Mapping Using Contact Fluorescent Imaging

3.2.1. Setting Up the Mapping System

1. Flow ethanol for 15 min then water for 15 min through the tubing and chamber to clean these components before mapping.

2. Turn on the temperature controller and set to 37°C. Allow a few minutes for the heater to warm up.

3. Flow Tyrode's solution and allow a few minutes for the solution to reach the chamber.

3.2.2. Voltage-Sensitive Dye Staining and Setting Up the Optics

1. Add 1 mL Tyrode's solution to dye aliquot for 10 µM final dye concentration.

2. Turn off Tyrode's solution flow from the pump and place the coverslip containing the cells into the chamber (see Note 40).

3. Aspirate out Tyrode's solution and quickly add the dye solution (from the edge and without allowing the cells to dry, to prevent damage). If the total volume is not enough to completely cover the coverslip, more dye solution is needed.

4. Stain for 10 min (see Note 41). Wash out the dye by turning the pump back on and superfusing the cells with warm Tyrode's solution prior to mapping.

5. Turn on data acquisition system and software.

6. Position the optical fiber bundle directly beneath the red emission filter, being careful to not crack the thin filter.

7. Position the LED excitation light directly above the chamber.

3.2.3. CFI Recording

1. Allow the cell preparation to reach steady-state following dye washout, that is, returning to a stable temperature, stable solution flow rate, etc.

2. Acquire recordings based on the desired protocol (see Note 42): During each recording, the data acquisition computer turns on the LED excitation light. The emitted fluorescent light passes through the red emission filter, is collected by the optical fiber bundle and is passed at the other end to the PDA. The PDA instrumentation converts the fluorescent light into analog signals, which are then filtered and amplified. The processed signals are digitized by the data acquisition computer and saved for postexperiment analysis.

3. Data files are processed and analyzed using custom MATLAB scripts (see Notes 43 and 44, Fig. 2).

3.3. Calcium Mapping Using a Single Lens and a CCD Camera

1. Clean and prepare the tubing and chamber as described for CFI (see Subheading 3.2.1).

3.3.1. Setting Up the Mapping System

1. Mount the lens in front of the CCD camera and position the camera above the experimental chamber. Mount the emission filter in front of the lens.

3.3.2. Calcium-Sensitive Dye Staining and Setting Up the Optics

2. Image a grid (with known spacing) in place of the coverslip to ensure the lens is positioned and focused properly on the grid and to calibrate the field of view (necessary for postexperiment analysis).

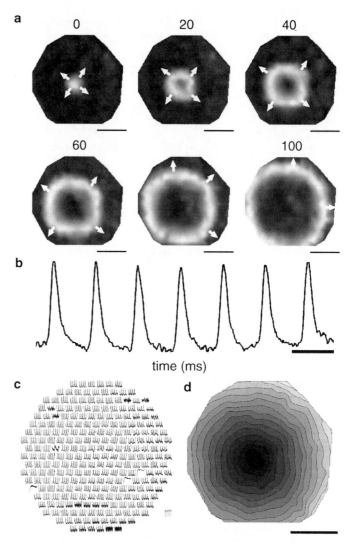

Fig. 2. Example of optical mapping of voltage using CFI. The mESC-CM monolayer was paced at 5 Hz by point stimulation near the center, and an electrical wave propagated out radially from the stimulus site. (**a**) Normalized voltage maps. Signals from individual recording sites were temporally filtered using a five-point median filter, detrended by subtraction of a fitted third-order polynomial, and normalized from 0 (*black*, resting potential) to 1 (*white*, peak depolarization potential). Maps were created by interpolating to a 100 × 100 μm grid over the mapping area. Channels with poor signal were not used for the interpolation. Scale bar is 5 mm. (**b**) A representative filtered and normalized signal from a single recording site. Scale bar is 200 ms. (**c**) Signal traces from the nominally 253 recording sites. (**d**) Isochrone map of activation times. Isochronal lines are spaced 10 ms apart. Scale bar is 5 mm.

3. Warm the dye stock solution up to room temperature for at least 30 min. Add 5 μL Pluronic stock to 5 μL dye stock solution. Add 1 mL Tyrode's solution and mix to obtain a 5 μM final dye concentration.

4. Pour the dye solution into a 30 mm diameter culture dish and quickly place the coverslip containing the cells in the dye solution. Cover the culture dish with aluminum foil and keep in the dark at room temperature for 20 min.

5. Aspirate out the dye solution and wash the coverslip with PBS three times.

6. Fill the chamber with warm Tyrode's solution and place the coverslip containing the cells into the chamber (see Note 40).

7. Superfuse the cells with warm Tyrode's solution for an additional 20 min prior to mapping (see Note 45).

8. Position the QTH lamp light guide such that light bathes the entire coverslip with the maximum possible intensity.

3.3.3. CCD Recording

1. Acquire recordings based on the desired protocol: During each recording, the user or computer turns on the QTH lamp excitation light by using a shutter. The emitted fluorescent light passes through the bandpass emission filter and is collected by the lens, mounted on the CCD camera. The CCD images are saved for postexperiment analysis.

2. Data files are processed and analyzed using custom MATLAB scripts (see Note 46, Fig. 3).

4. Notes

1. Currently there are at least four companies producing MEA systems commercially, including Alpha Med Sciences (Berkeley, CA), Multichannel Systems (Reutlingen, Germany), Multi Micro Electrode Systems LLC (Bellflower, CA), and Plexon (Dallas, TX). The Multichannel Systems MEA has been the one most widely used for recordings of cardiomyocytes and SC-CMs, and an application note is available on their website specifically detailing MEA mapping of human embryonic SC-CMs (31).

2. Typical light sources are xenon arc (XA) lamps, quartz-tungsten halogen (QTH) lamps, light-emitting diodes (LEDs), and lasers. XA and QTH lamps are high power stable light sources but are typically much more expensive than LEDs. Lasers provide high intensity light only at a single wavelength, and are noisier compared with lamp sources. Other considerations regarding excitation sources, such as power, noise, and cost, have been reviewed in more detail (23, 32, 33). XA and QTH lamps are available from Olympus (Orangeburg, NY), Oriel (Newport Corporation, Irvine, CA), and Osram Sylvania (Danvers, MA). LEDs are an inex-

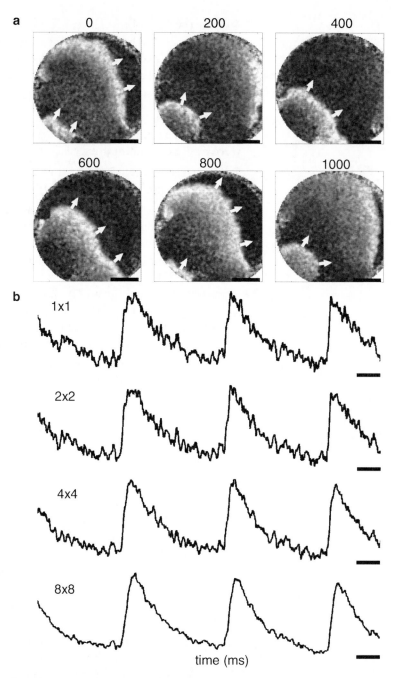

Fig. 3. Example of optical mapping of calcium using a CCD camera. A spontaneous, slow moving wave of calcium originated near the coverslip edge and propagated across the mESC-CM monolayer. The mapping area was 20.5 × 20.5 mm, with a 160 × 160 µm pixel size. Images were acquired at 490 Hz frame rate. Camera image falling outside the 21 mm diameter circular coverslip is not shown. (**a**) Normalized calcium maps. Individual pixels were spatially filtered using a 3 × 3 box filter, temporally filtered using a five-point median filter, detrended by subtraction of a third-order polynomial fit, and normalized from 0 (*black*, resting calcium level) to 1 (*white*, peak calcium level). Scale bar is 5 mm. (**b**) Individual normalized traces from a single site at different levels of software binning. As binning size increases, the signal-to-noise ratio of the traces improves. Scale bar is 200 ms.

pensive light source and are commercially available from a variety of companies, including Fairchild Semiconductor (South Portland, ME), Kingbright (Taipei, Taiwan), and Vishay Semiconductor (Malvern, PA). High power LEDs are available from Lumileds (San Jose, CA). For microscopic mapping, a standard tungsten, mercury arc, or xenon arc lamp accessory can provide the necessary excitation light. At high magnification, too high a lamp intensity may cause rapid photobleaching and phototoxicity, so that a neutral density filter may be needed to attenuate the light output.

3. An optical bandpass filter is usually placed in front of the light source to pass only light that falls within the excitation spectrum. Since lasers are single wavelength sources, an excitation filter is unnecessary.

4. For optical mapping of cardiac tissues and monolayers, the photodetectors used include PDAs, CCD cameras, and complementary metal-oxide semiconductor (CMOS) cameras. PDAs can provide high time resolution and sensitivity, owing to a large pixel size, but typically poor spatial resolution. CCD cameras typically provide a high spatial resolution compared with a PDA but at a lower imaging frame rate. Time resolution can be improved by pixel binning although this sacrifices the spatial resolution of the system. CMOS cameras are becoming more popular in use for cardiac optical mapping, as they provide both high spatial and temporal resolution but with some loss of sensitivity. A CCD or CMOS camera requires a light-collecting lens and is therefore not practical for contact imaging. Commercially available PDAs are available from Hamamatsu (Japan) and Redshirt Imaging (Decatur, GA). CCD cameras are available from Andor (South Windsor, CT), Photometrics (Roper Scientific GmbH, Germany), Redshirt Imaging, and SciMedia (Costa Mesa, CA). CMOS cameras are available from Redshirt and SciMedia. Electron multiplication CCD (EMCCD) cameras are a recent improvement that provides increased sensitivity and are available from Andor, Photometrics and several other companies. Additional information regarding photodetectors for optical mapping can be found in recent reviews (15, 23, 33).

5. Alternatively, the optical fiber bundle can be eliminated and a PDA can be placed directly underneath and in contact with the emission filter, at the risk of damage from solution spill. Or, a lens can substitute for the bundle, with the loss of some light collection.

6. The sample is placed at the focal plane of one of the lenses, the photodetector is placed at the focal plane of a second lens in tandem with the first, and the two lenses face each other (34). The ratio of the focal lengths determines the magnifica-

tion of the system. Modifying the focal lengths of the lens(es) and/or their positions in the single lens or TL configurations enables a change of the magnification and therefore spatial resolution of the system, whereas the CFI configuration is fixed at 1:1 magnification.

7. Low magnification objectives (10× or less) generally collect much less light across the field of view than do high magnification objectives because of their much lower numerical aperture. Spatial and/or temporal averaging may be necessary to improve signal quality.

8. Polydimethylsiloxane (PDMS) or glass rings also work well as barriers. This is an important step when working with a limited number of cells, which is often the case when studying SC-CMs.

9. It is important not to shake or jar the fibronectin stock solution, as this will cause polymerization and reduce cell adherence. When diluting, always add the fibronectin stock solution drop by drop to cold PBS.

10. In general, cell growth and phenotype depends substantially on the culture environment, including, in particular, composition of the culture media. When changing media components (including the lot number), screening of cell viability and other cell characteristics should be performed.

11. MEA recording chambers/amplifiers are also available to accommodate upright microscopes.

12. An antivibration table is optional but its use greatly decreases mechanically generated noise.

13. Tyrode's solution as used in the literature consists of 1.8 mM $CaCl_2$, 135–140 mM NaCl, 5.4 mM KCl, 1 or 2 mM $MgCl_2$, 5 or 10 mM glucose, 5 or 10 mM HEPES, and 0–5.5 NaH_2PO_4. For our mouse embryonic SC-CMs, we have found 1.8 mM $CaCl_2$, 135 mM NaCl, 5.4 mM KCl, 2 mM $MgCl_2$, 10 mM glucose, and 10 mM HEPES to be well tolerated. A 5× stock solution without the $CaCl_2$ and glucose is prepared, brought to pH 7.4 with NaOH at 37°C, and then stored at 4°C for up to several weeks. When ready to use, dilute the stock to 1×, add the $CaCl_2$ and glucose, warm the Tyrode's solution to 37°C, and recheck the pH. Final osmolality of the solution should be approximately 293 mOsm/kg H_2O.

14. Experimental chambers suitable for optical mapping require certain mechanical and optical properties. The chamber needs to provide a stable environment for the cell monolayer. It is desirable for the chamber to be maintained at a constant temperature and allow for superfusion of both control and drug solutions. If solutions are flowed, a lid will help to prevent

vibration of the solution surface from affecting the fluorescent signals. Additionally, the floor and lid of the chamber must be made of materials that are optically transparent and do not autofluoresce.

15. A flow system enables continuous superfusion of the cells and switching of control Tyrode's solution with a drug solution. It also produces a more uniform temperature distribution across the chamber.

16. The pump should have a large number of rollers to minimize fluctuations in flow rate. The flow rate can be adjusted for the particular chamber being used.

17. A heating system is necessary to carry out experiments at physiological temperatures. Either the solution flowing into the chamber can be heated, or the platform supporting the chamber can be heated, or both. The solution can be heated by a water bath (Cole-Parmer) running through a water jacket or by an electronic heating unit placed in series with the tubing. Commercially available temperature controllers, heated platforms, and enclosures are available from Cole-Parmer, Omega Engineering (Stamford, CT), and Warner Instruments (Hamden, CT).

18. Platinum wire, silver chloride wire or salt bridges are typically used for electrodes. These electrode materials are stable (particularly platinum) and nontoxic to the cells. However, excessive levels of current can cause electrolytic bubbling, which can interfere with the optical recording and also damage the cells.

19. Commercially available stimulators are also available from AM-Systems (Sequim, WA), and World Precision Instruments (Sarasota, FL), to name a few.

20. Other commonly used voltage-sensitive dyes are di-8-ANEPPS and RH-237 (33) (Invitrogen, Carlsbad, CA).

21. We typically aliquot the voltage-sensitive dye stock solution into centrifuge tubes sufficient for 1 week of experiments. Before aliquoting, the dye stock solution should be warmed to approximately 37°C.

22. The wavelength of the excitation source depends on the excitation spectrum of the fluorescent dye. Fluorescent dyes are excited by a range of wavelengths, sometimes with multiple peaks. The excitation source should produce light near a peak of the excitation spectrum and no light in the emission spectrum of the dye. One peak of the di-4-ANEPPS spectrum is at the green wavelength of ~500 nm, and the other is at the blue wavelength of ~440 nm, when measured at a 610 nm emission wavelength (35). Another consideration is that for voltage-

sensitive dyes, the excitation and emission spectra of the dye can change significantly when the dye is bound to a cell membrane than when it is in solution (36). Therefore it is important to select the excitation wavelength with regard to the spectrum for the membrane-bound form of the dye.

23. As with the excitation source, the choice of emission filter depends on the emission spectrum of the fluorescent dye. The emission spectrum changes as a function of the excitation wavelength. When excited by green wavelength light (520–570 nm), di-4-ANEPPS emission has a single peak at the red wavelength of ~630 nm (as utilized in this chapter). However, when excited by blue wavelength light (440–490 nm), di-4-ANEPPS emission has two peaks, with one at the green wavelength of ~540 nm and the other at the red wavelength of ~630 nm. During membrane depolarization, green fluorescence increases and red fluorescence decreases (37). Dyes with dual peaks of either the excitation or emission spectrum can be used for ratiometric measurements to improve signal quality by reducing distortion owing to non-uniform dye uptake, non-uniform excitation illumination, variable optical path, and motion artifact (37) or to quantify the transmembrane voltage of nonexcitable cells (35, 38).

24. CFI requires the smallest possible thickness of the emission filter to minimize crosstalk between adjacent recording sites (14). Commercially available filters can be purchased from Chroma Technology, Edmund Optics (Barrington, NJ), Omega Optical (Brattleboro, VT), and Thorlabs (Newton, NJ), but their thickness is typically several millimeters and hence, is inadequate. Custom filters having 0.3 mm thickness are available but are quite expensive and prone to breakage.

25. For a 1 kHz sampling rate, the filtering stage includes a 350 Hz (can be up to 500 Hz) low pass cutoff filter to ensure that the Nyquist criteria is met and that signal distortion via aliasing is prevented. The gain stage is designed to utilize the full dynamic range of the data acquisition board of the mapping computer.

26. The data acquisition system can also be designed to coordinate the timing of the excitation source light. One advantage of LED light sources is that they can be rapidly turned on/off, with delays as short as microseconds; however, they will then be prone to baseline drift owing to heating effects. Arc and halogen lamps can be turned on/off with the addition of a digitally controlled mechanical shutter, which typically opens/closes on the order of several milliseconds.

27. To avoid signal distortion, the sampling rate should be set high enough to adequately capture the fastest event, which

for a cardiac monolayer is typically the action potential upstroke. In practice, however, the sampling rate may be limited by the maximum speed of the acquisition system.

28. Other calcium-sensitive dyes commonly used for cell culture include Indo-1, Fura-2, Fluo-3, and Fluo-4 (Invitrogen) (39). The advantages of the different dyes, as well as practical dual voltage and calcium imaging dye pairs, have been reviewed for cardiac tissue (15, 23, 24, 33).

29. Rhod-2 has an excitation peak at ~550 nm. Unlike Rhod-2, other calcium-sensitive dyes, such as Indo-1 (40) and Fura-2 (41), have shifts in the excitation or emission spectrum upon calcium binding, and this property enables those dyes to be used for ratiometric measurements of absolute intracellular calcium levels.

30. For many commercially available CCD cameras, image acquisition software is packaged together with the camera. The sampling rate is limited by the amount of time the camera shutter is open and the time required to store data. Pixel binning reduces the amount of data, and therefore storage time, enabling a faster sampling rate. With the Andor iXon+ 860 EMCCD at maximum 128×128 pixel resolution, images are acquired at a maximum of ~490 Hz. When higher temporal resolution is deemed necessary, 2×2 hardware binning enables acquisition at ~860 Hz, and binning up to 8×8 can be used to maximize the frame rate at ~2 kHz. If higher temporal resolution is not necessary, the sampling rate can be reduced, which allows for a longer shutter open time and improved signal-to-noise ratio.

31. An ideal lens has a high numerical aperture for maximal light gathering and projection onto the CCD chip. For macroscopic mapping, demagnification lenses are required, and C-mount lenses that attach directly to the camera are commercially available from Linos Photonics (Goettingen, Germany), Navitar (Rochester, NY), Nikkor (Nikon, Melville, NY), and Schneider Optics (Germany). Additionally, the use of a camera enables images to be taken to document structural information, such as cell morphology and orientation, and cells and cell regions that are identified by fluorescent labels.

32. Rhod-2 emission has a single peak at the yellow wavelength of ~580 nm.

33. The MEA should be coated with extracellular matrix protein such as fibronectin to maximize SC-CM adhesion. Other common choices are laminin and collagen. Cells tend to have more difficulty adhering to new MEAs or MEAs that have been in storage, because the surface of the MEAs tends to be hydrophobic. Multi Channel Systems gives several suggestions to improve cell adhesion to their MEAs,

including plasma cleaning, protein coating, and preculturing (42). Do not treat the chamber with anything that might damage the electrodes or impede their ability to contact the cells.

34. Sufficient humidity is necessary to keep the culture medium from evaporating, which would change its osmolarity and ion concentrations. Placing the entire MEA in a sterile Petri dish of sterile water prior to placing it in the incubator is recommended by other users.

35. A small, sterile Petri dish or a gas permeable/vapor impermeable membrane can also be used as a lid (2, 43).

36. In our experience, mouse embryonic SC-CMs in Tyrode's solution can be irreparably damaged if placed in a 5% CO_2 incubator.

37. If using Cell Tracker or other fluorescent markers to identify SC-CMs within a coculture, acquire a fluorescence image for later localization of the cells.

38. During recording, apply desired perturbations to the system, for example, ion channel blockers, decoupling agents, or other pharmacological agents. If using pacing chambers, pacing of cells on the MEA is possible. When electrically pacing cells on MEAs, one major challenge is the stimulus artifact. Multi Channel System recommends using MEAs with larger stimulation electrodes for pacing cardiomyocytes and a special stimulation adapter when pacing cardiac tissue. If multiple time points are desired, after finishing each recording aspirate out the Tyrode's solution and replace with culture medium containing antibiotics. This will minimize the chance that contamination develops. Each day, observe the cells and culture medium carefully under the microscope for any signs of contamination.

39. Before analysis, field potentials (FPs) from each recording electrode may be temporally filtered in hardware before digitization or in software afterward. Typical measurements when assessing MEA recordings include: rhythmicity, conduction, and identification of the origin and route of electrical propagation. To assess propagation, the activation times must be computed at each recording electrode. For the voltage V_i at recording electrode i, the activation time can be computed as the time of the maximal negative first derivative of the signal (dV/dt_{min}). Rhythmicity can be measured by variability in the cycle length, computed from the difference of successive activation times. Repolarization times have also been measured from MEA recordings from the timing of a second, smaller slow deflection (following the first deflection for activation) (44) or return to baseline (6). Field potential duration is then measured as the difference between activation and repolariza-

tion times. However, since MEA recordings are only in the microvolt range for the larger activation deflection, measurement of the repolarization time will be difficult and imprecise. Conduction velocity (CV) can be measured by taking the distance of a path perpendicular to the direction of electrical propagation, and dividing by the difference of activation times at the path endpoints.

40. Movement of the cell monolayer owing to cell contraction can cause distortions in the acquired optical signals. To minimize motion artifact, either the cells must be attached to a rigid substrate (for example, glass or plastic coverslip) or an excitation-contraction uncoupler must be used if the cell monolayer is unattached or grown on a compliant hydrogel or polymer scaffold. When optically mapping cells grown on polymer scaffolds, we found blebbistatin (10 μM) (45) to yield the best results compared with alternative options of cytochalasin D (46) or 2,3-Butanedione monoxime (BDM) (47). In addition, when optically mapping cells cultured on scaffolds, the scaffold must be immobilized, for example, through the use of pins or a restraining membrane (48), particularly when flowing solution through the mapping chamber.

41. Cell staining time for cardiomyocytes enzymatically dissociated from native cardiac tissue is typically shorter, ~5 min.

42. The total excitation light exposure time should be minimized, because excited dye produces small amounts of toxic by-products (49). The death of individual cells will reduce regional optical signals and ultimately create large regions of conduction block in the cell monolayer. Even with minimal light exposure time, solution flow will also reduce signal amplitude as a function of time (50).

43. During each recording, the fluorescence baseline decreases owing to photobleaching of the dye (and heating of the light source if it is switched on and off, as may be the case with LEDs). For data analysis, the baseline drift can be compensated in hardware by a sample-and-hold circuit on each channel or in software by subtraction of a linear, polynomial, or exponential fit from the signal at each recording site. Hardware elimination involves the addition of more electronic circuitry but eliminates the need for greater than 8-bit resolution in the analog-to-digital conversion, which reduces the cost and increases the speed of the mapping system.

44. The fluorescence signal from each CFI recording site is typically digitally processed before analysis. Processing stages can include temporal filtering (such as elliptical, Butterworth, or median filter types), baseline detrending (see Note 43), and normalization. Two important data analysis measurements are the conduction velocity (CV) and action potential

duration (APD), and their rate-dependence (51) and uniformity across the monolayer. As already described for MEA recordings, to assess propagation of electrical activity, the activation times must be determined at each recording site. For fluorescence voltage signal $F_i(t)$ at site i, the activation time can be computed as the time of the maximum first derivative of the action potential upstroke, dF_i/dt_{max}. Alternatively, the activation time can be computed as the time of the midpoint (50% amplitude) of the action potential upstroke. Repolarization time can be computed as the time of 80% recovery from the peak amplitude of the action potential, or the times for 50 or 90% repolarization may be used. Alternatively, the repolarization time can be computed as the time of the maximum second derivation of the downstroke, d^2F_i/dt^2, although in practice, this is rarely done. APD is measured as the difference between activation and repolarization times. CV can be computed as described for MEA recordings (see Note 39).

45. Rhod-2-AM, the membrane permeable form of Rhod-2, does not bind calcium. The 20 min incubation time period allows endogenous intracellular esterases to hydrolyze Rhod-2-AM to Rhod-2. Once this occurs, the dye is no longer membrane permeable and remains trapped inside the cell.

46. As described for CFI recordings (see Note 44), the fluorescence signal at each CCD pixel is typically digitally processed before analysis. Processing stages can include those previously described (temporal filtering, baseline detrending, normalization), as well as spatial binning or filtering (such as box or Gaussian filter types). Activation time and conduction velocity can be computed for calcium waves as previously described for CFI voltage signals. Another important measurement of a calcium signal is the relaxation time (RT) or time constant. RT can be computed as the time of 50 or 80% recovery from the peak amplitude of the calcium transient. The relaxation time constant can be computed by fitting a single exponential to the calcium transient decay.

Acknowledgments

This work was supported by NIH grants R01 HL066239 (L.T.) and T32-HL07581 (A. Shoukas), and grants from the Joint Technion-Hopkins Program for the Biomedical Sciences and Biomedical Engineering (L.T. and L. Gepstein) and from the Maryland Stem Cell Research Fund (L.T.). We thank Dr. Lior Gepstein for training E.L. in his lab on the use of MEAs.

References

1. Reppel, M., Pillekamp, F., Lu, Z. J., Halbach, M., Brockmeier, K., Fleischmann, B. K., and Hescheler, J. (2004) Microelectrode arrays: a new tool to measure embryonic heart activity. *J Electrocardiol* 37 Suppl, 104–9.

2. Egert, U., and Meyer, T. (2005) Heart on a chip – extracellular multielectrode recordings from cardiac myocytes in vitro. in *Practical Methods in Cardiovascular Research* (Dhein, S., Mohr, F. W., and Delmar, M., Eds.) pp 432–453, Springer, Berlin.

3. Binah, O., Dolnikov, K., Sadan, O., Shilkrut, M., Zeevi-Levin, N., Amit, M., Danon, A., and Itskovitz-Eldor, J. (2007) Functional and developmental properties of human embryonic stem cells-derived cardiomyocytes *J Electrocardiol* 40, S192–6.

4. Igelmund, P., Fleischmann, B. K., Fischer, I. R., Soest, J., Gryshchenko, O., Bohm-Pinger, M. M., Sauer, H., Liu, Q., and Hescheler, J. (1999) Action potential propagation failures in long-term recordings from embryonic stem cell-derived cardiomyocytes in tissue culture *Pflugers Arch* 437, 669–79.

5. Kehat, I., Kenyagin-Karsenti, D., Snir, M., Segev, H., Amit, M., Gepstein, A., Livne, E., Binah, O., Itskovitz-Eldor, J., and Gepstein, L. (2001) Human embryonic stem cells can differentiate into myocytes with structural and functional properties of cardiomyocytes *J Clin Invest* 108, 407–14.

6. Banach, K., Halbach, M. D., Hu, P., Hescheler, J., and Egert, U. (2003) Development of electrical activity in cardiac myocyte aggregates derived from mouse embryonic stem cells *Am J Physiol Heart Circ Physiol* 284, H2114–23.

7. Beeres, S. L., Atsma, D. E., van der Laarse, A., Pijnappels, D. A., van Tuyn, J., Fibbe, W. E., de Vries, A. A., Ypey, D. L., van der Wall, E. E., and Schalij, M. J. (2005) Human adult bone marrow mesenchymal stem cells repair experimental conduction block in rat cardiomyocyte cultures *J Am Coll Cardiol* 46, 1943–52.

8. Kehat, I., Khimovich, L., Caspi, O., Gepstein, A., Shofti, R., Arbel, G., Huber, I., Satin, J., Itskovitz-Eldor, J., and Gepstein, L. (2004) Electromechanical integration of cardiomyocytes derived from human embryonic stem cells *Nat Biotechnol* 22, 1282–9.

9. Caspi, O., Itzhaki, I., Arbel, G., Kehat, I., Gepstien, A., Huber, I., Satin, J., and Gepstein, L. (2009) In vitro electrophysiological drug testing using human embryonic stem cell derived cardiomyocytes. *Stem Cells Dev* 18, 161-72.

10. Meyer, T., Boven, K. H., Gunther, E., and Fejtl, M. (2004) Micro-electrode arrays in cardiac safety pharmacology: a novel tool to study QT interval prolongation *Drug Saf* 27, 763–72.

11. Windisch, H., Ahammer, H., Schaffer, P., Muller, W., and Platzer, D. (1995) Optical multisite monitoring of cell excitation phenomena in isolated cardiomyocytes *Pflugers Arch* 430, 508–18.

12. Rohr, S., and Salzberg, B. M. (1994) Multiple site optical recording of transmembrane voltage (MSORTV) in patterned growth heart cell cultures: assessing electrical behavior, with microsecond resolution, on a cellular and subcellular scale *Biophys J* 67, 1301–15.

13. Fast, V. G., and Kleber, A. G. (1993) Microscopic conduction in cultured strands of neonatal rat heart cells measured with voltage-sensitive dyes *Circ Res* 73, 914–25.

14. Entcheva, E., Lu, S. N., Troppman, R. H., Sharma, V., and Tung, L. (2000) Contact fluorescence imaging of reentry in monolayers of cultured neonatal rat ventricular myocytes *J Cardiovasc Electrophysiol* 11, 665–76.

15. Efimov, I. R., Nikolski, V. P., and Salama, G. (2004) Optical imaging of the heart *Circ Res* 95, 21–33.

16. Salama, G., and Morad, M. (1976) Merocyanine 540 as an optical probe of transmembrane electrical activity in the heart. *Science* 191, 485–7.

17. Gray, R. A., Jalife, J., Panfilov, A., Baxter, W. T., Cabo, C., Davidenko, J. M., and Pertsov, A. M. (1995) Nonstationary vortexlike reentrant activity as a mechanism of polymorphic ventricular tachycardia in the isolated rabbit heart *Circulation* 91, 2454–69.

18. Dillon, S. M. (1991) Optical recordings in the rabbit heart show that defibrillation strength shocks prolong the duration of depolarization and the refractory period *Circ Res* 69, 842–56.

19. Morad, M., and Salama, G. (1979) Optical probes of membrane potential in heart muscle *J Physiol* 292, 267–95.

20. Sato, D., Shiferaw, Y., Garfinkel, A., Weiss, J. N., Qu, Z., and Karma, A. (2006) Spatially discordant alternans in cardiac tissue: role of calcium cycling *Circ Res* 99, 520–7.

21. Hwang, S. M., Yea, K. H., and Lee, K. J. (2004) Regular and alternant spiral waves of contractile motion on rat ventricle cell cultures *Phys Rev Lett* 92, 198103.

22. Tung, L., and Zhang, Y. (2006) Optical imaging of arrhythmias in tissue culture *J Electrocardiol* 39, S2–6.

23. Entcheva, E., and Bien, H. (2006) Macroscopic optical mapping of excitation in cardiac cell networks with ultra-high spatiotemporal resolution *Prog Biophys Mol Biol* 92, 232–57.

24. Fast, V. G. (2005) Recording action potentials using voltage-sensitive dyes. in *Practical Methods in Cardiovascular Research* (Dhein, S., Mohr, F. W., and Delmar, M., Eds.) pp 233–255, Springer, Berlin.

25. Lagostena, L., Avitabile, D., De Falco, E., Orlandi, A., Grassi, F., Iachininoto, M. G., Ragone, G., Fucile, S., Pompilio, G., Eusebi, F., Pesce, M., and Capogrossi, M. C. (2005) Electrophysiological properties of mouse bone marrow c-kit+ cells co-cultured onto neonatal cardiac myocytes *Cardiovasc Res* 66, 482–92.

26. Orlandi, A., Pagani, F., Avitabile, D., Bonanno, G., Scambia, G., Vigna, E., Grassi, F., Eusebi, F., Fucile, S., Pesce, M., and Capogrossi, M. C. (2008) Functional properties of cells obtained from human cord blood CD34+ stem cells and mouse cardiac myocytes in coculture *Am J Physiol Heart Circ Physiol* 294, H1541–9.

27. Dolnikov, K., Shilkrut, M., Zeevi-Levin, N., Gerecht-Nir, S., Amit, M., Danon, A., Itskovitz-Eldor, J., and Binah, O. (2006) Functional properties of human embryonic stem cell-derived cardiomyocytes: intracellular Ca²⁺ handling and the role of sarcoplasmic reticulum in the contraction *Stem Cells* 24, 236–45.

28. Kapur, N., Mignery, G. A., and Banach, K. (2007) Cell cycle-dependent calcium oscillations in mouse embryonic stem cells *Am J Physiol Cell Physiol* 292, C1510–8.

29. Sauer, H., Hofmann, C., Wartenberg, M., Wobus, A. M., and Hescheler, J. (1998) Spontaneous calcium oscillations in embryonic stem cell-derived primitive endodermal cells *Exp Cell Res* 238, 13–22.

30. Egert, U., Knott, T., Schwarz, C., Nawrot, M., Brandt, A., Rotter, S., and Diesmann, M. (2002) MEA-Tools: an open source toolbox for the analysis of multi-electrode data with MATLAB *J Neurosci Methods* 117, 33–42.

31. Multi Channel Systems (2006) MEA Application Note: Human Embryonic Stem Cell Derived Cardiac Myocytes (hESC-CM). Multi Channel Systems MCS GmbH.

32. Fast, V. G. (2005) Simultaneous optical imaging of membrane potential and intracellular calcium *J Electrocardiol* 38, 107–12.

33. Tritthart, H. A. (2005) Optical techniques for the recording of action potentials. in *Practical Methods in Cardiovascular Research* (Dhein, S.,

Mohr, F. W., and Delmar, M., Eds.) pp 215–232, Springer, Berlin.

34. Ratzlaff, E. H., and Grinvald, A. (1991) A tandem-lens epifluorescence macroscope: hundred-fold brightness advantage for wide-field imaging *J Neurosci Methods* 36, 127–37.

35. Montana, V., Farkas, D. L., and Loew, L. M. (1989) Dual-wavelength ratiometric fluorescence measurements of membrane potential *Biochemistry* 28, 4536–9.

36. Muller, W., Windisch, H., and Tritthart, H. A. (1986) Fluorescent styryl dyes applied as fast optical probes of cardiac action potential *Eur Biophys J* 14, 103–11.

37. Knisley, S. B., Justice, R. K., Kong, W., and Johnson, P. L. (2000) Ratiometry of transmembrane voltage-sensitive fluorescent dye emission in hearts *Am J Physiol Heart Circ Physiol* 279, H1421–33.

38. Beach, J. M., McGahren, E. D., Xia, J., and Duling, B. R. (1996) Ratiometric measurement of endothelial depolarization in arterioles with a potential-sensitive dye *Am J Physiol* 270, H2216–27.

39. Takahashi, A., Camacho, P., Lechleiter, J. D., and Herman, B. (1999) Measurement of intracellular calcium *Physiol Rev* 79, 1089–125.

40. Katra, R. P., Pruvot, E., and Laurita, K. R. (2004) Intracellular calcium handling heterogeneities in intact guinea pig hearts *Am J Physiol Heart Circ Physiol* 286, H648–56.

41. Field, M. L., Azzawi, A., Styles, P., Henderson, C., Seymour, A. M., and Radda, G. K. (1994) Intracellular Ca²⁺ transients in isolated perfused rat heart: measurement using the fluorescent indicator Fura-2/AM. *Cell Calcium* 16, 87–100.

42. Multi Channel Systems (2005) Microelectrode Array (MEA) User Manual. Multi Channel Systems MCS GmbH.

43. Potter, S. M., and DeMarse, T. B. (2001) A new approach to neural cell culture for long-term studies *J Neurosci Methods* 110, 17–24.

44. Yamamoto, M., Honjo, H., Niwa, R., and Kodama, I. (1998) Low-frequency extracellular potentials recorded from the sinoatrial node *Cardiovasc Res* 39, 360–72.

45. Fedorov, V. V., Lozinsky, I. T., Sosunov, E. A., Anyukhovsky, E. P., Rosen, M. R., Balke, C. W., and Efimov, I. R. (2007) Application of blebbistatin as an excitation-contraction uncoupler for electrophysiologic study of rat and rabbit hearts *Heart Rhythm* 4, 619–26.

46. Wu, J., Biermann, M., Rubart, M., and Zipes, D. P. (1998) Cytochalasin D as excitation-contraction uncoupler for optically mapping action

potentials in wedges of ventricular myocardium *J Cardiovasc Electrophysiol* 9, 1336–47.

47. Cheng, Y., Mowrey, K., Efimov, I. R., Van Wagoner, D. R., Tchou, P. J., and Mazgalev, T. N. (1997) Effects of 2,3-butanedione monoxime on atrial-atrioventricular nodal conduction in isolated rabbit heart. *J Cardiovasc Electrophysiol* 8, 790–802.

48. Bursac, N., Loo, Y., Leong, K., and Tung, L. (2007) Novel anisotropic engineered cardiac tissues: studies of electrical propagation *Biochem Biophys Res Commun* 361, 847–53.

49. Schaffer, P., Ahammer, H., Muller, W., Koidl, B., and Windisch, H. (1994) Di-4-ANEPPS causes photodynamic damage to isolated cardiomyocytes *Pflugers Arch* 426, 548–51.

50. Windisch, H., Muller, W., and Tritthart, H. A. (1985) Fluorescence monitoring of rapid changes in membrane potential in heart muscle *Biophys J* 48, 877–84.

51. Boyett, M. R., and Jewell, B. R. (1980) Analysis of the effects of changes in rate and rhythm upon electrical activity in the heart *Prog Biophys Mol Biol* 36, 1–52.

Chapter 15

Assessment of Cardiac Conduction: Basic Principles of Optical Mapping

Chunhua Ding and Thomas H. Everett IV

Abstract

Extracellular recordings acquired from electrodes placed on the surface of cardiac tissue have traditionally been used to study the electrophysiological properties of the tissue. While this technique has been used in several studies that have increased our understanding of cardiac arrhythmias and action potential propagation, there are several limitations that have prevented us from seeing a bigger picture of arrhythmia mechanisms. These limitations include the limited number of electrodes and unstable recordings. Optical mapping was developed to increase the temporal and spatial resolution over traditional electrode recordings and ultimately the accuracy of the data analysis. This technology involves using a voltage-sensitive dye that binds to the cell membrane. The fluorescence changes of the dye have a linear relationship to the action potential changes of the cell membrane. These fluorescent changes can then be detected by a photodiode array, a CCD camera or a CMOS camera. This will allow the recording of the action potential in hundreds to thousands of different sites simultaneously. Presented in this chapter are the materials and hardware needed along with step-by-step instructions on setup and techniques used in optical mapping for larger tissue preparations.

Key words: Optical mapping, Conduction, Electrophysiology

1. Introduction

In order to study cardiac arrhythmias and to gain insights as to why certain arrhythmias can occur in diseased hearts, there are two main areas of concentration. One is to study the electrophysiological and conduction properties of the cardiac tissue in order to understand how an action potential is traveling through the tissue; another is to study the mechanism of the resulting arrhythmia, which includes initiation and propagation. In order to study these two main areas, traditionally, electrodes have been placed on the surface of the tissue to record extracellular recordings.

Randall J. Lee (ed.), *Stem Cells for Myocardial Regeneration: Methods and Protocols*, Methods in Molecular Biology, vol. 660,
DOI 10.1007/978-1-60761-705-1_15, © Springer Science+Business Media, LLC 2010

For cardiac tissue, in whole heart in vivo experiments, these electrodes can be placed either inside the heart on the endocardial surface, or on the outside of the heart on the epicardial surface or both simultaneously (1–3). These techniques have several limitations which include the number of electrodes that can be placed on the surface of the tissue, and the stability of the recordings. In addition, when looking at electrical activation, the reliability of the results has come into question and in some cases the data is difficult to interpret (4). Optical mapping provides a recording of the activation and repolarization of the cardiac tissue from several sites simultaneously which provides an increase in the temporal and spatial resolution over traditional electrode recordings. As shown in Fig. 1., this technique of mapping involves a light source, a series of filters and optics, and a photodetector. Optical mapping involves injecting a voltage-sensitive dye into the tissue preparation. The voltage-sensitive dye then binds to the cell membrane and emits a certain wavelength of fluorescence depending on its emission spectra. The fluorescence changes of the dye have a linear relationship to the cell membrane voltage changes. The fluorescence is then detected by a photodiode array, a CCD or a CMOS camera. Because the intensity of the fluorescence is related only to the cell membrane voltage and

Optical mapping system for large tissue preparations

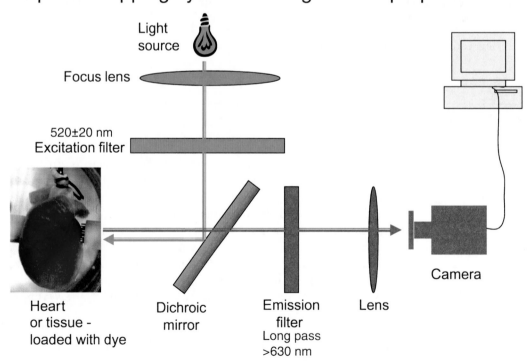

Fig. 1. Schematic of optical mapping system.

not to any other electrical phenomena, optical mapping is a valuable tool to study cardiac fibrillation and defibrillation with no simulation pulse artifacts.

Due to the change in membrane potential, the voltage is on the order of milliseconds, voltage sensitive dyes must be able to follow the voltage changes accordingly. The dyes that are used in optical mapping are considered fast dyes that have response times capable of following the membrane potential changes that occur with an action potential. Two dyes that fit this criteria are di-4-ANEPPS and PGH-1 (Pittsburgh 1) (5). These dyes show a linear change in fluorescence with changes in membrane potential.

Once the dye has been administered, and the preparation is still viable, a means of uncoupling the excitation/contraction process is needed to avoid motion artifacts which can distort the optical recordings from beating tissue. A pharmacologic approach is commonly used with the use of 2,3-butanedione monoxime which is a noncompetitive inhibitor of muscle myosin II, or cytochalasin D which impairs actin filament polymerization (6). Recently, a new agent has been developed, blebbistatin, which inhibits adenosine triphosphatases of muscle myosin II (7). Care must be taken in using these agents as 2,3-butanedione monoxime and cytochalasin D have been shown to affect intracellular Ca handling, ion channel kinetics, and ultimately the characteristics of the action potential (6, 8–11). Blebbistatin has been shown to have no effect on the action potential or intracellular calcium transients in rabbit cardiac tissue (7). However, more testing with this agent needs to be performed. An adequately perfused paralyzed preparation, with voltage sensitive dye loaded is then ready for the experimental protocol.

This chapter will focus on optical mapping of large tissue preparations. For optical mapping of cells and cell cultures, please see the other chapter 14 within this section.

2. Materials

2.1. Optical Mapping System Setup

1. PC Computer with a signal acquisition card and a LCD flat panel monitor (see Note 1).

2. Fluorescence detector: It could be either a Photodiode array (256-element, 16×16 array photodiode camera – C4675, Hamamatsu, Bridgewater, New Jersey, USA) (12), CCD camera (Model: MiCAM02, SciMedia USA Ltd.) or CMOS camera. (Model: MiCAM ULTIMA, SciMedia USA Ltd.)

3. Excitation filter: To filter light for excitation (green light 520 ± 20 nm) (Omega Optical, Inc. VT, USA or Chroma Technology Corp. VT, USA).

4. Emission filter: To acquire the right wavelength of fluorescence for recording (red light, long pass > 630 nm) (Omega Optical, Inc. VT, USA or Chroma Technology Corp. VT, USA).

5. Dichroic mirrors: 520–560 nm reflect (>90%), >600 nm (>85%) pass.

6. Lens: Depends on the detection device and size of the viewing area. For a larger tissue preparation such as canine and swine, the recording area is 4 cm² with the photodiode array using a 50 mM-focal length lens (Diameter 72 mM, F-Stop 1.4, Canon); for the CCD or CMOS cameras, the recording area is greater and a 25-mM focal length lens is used (Navitar DO-2595, Diameter 40.5 mM, F-Stop 0.95. Navitar Inc. NY, USA).

7. Plunge electrodes: Used to record electrical signals from the preparation. Make sure that the wire coating is stripped away at the connection sites. (0.03 mM MP35N wire (Fort Wayne Metals, Fort Wayne Indiana, USA) (13)).

8. Amplifier: For gaining and filtering signals obtained from the plunge electrodes that are inserted into the preparation (World Precision Instruments – WPI Inc., at least 2 channels).

9. Power supply: Various models depending on the lamps that are used.

10. Excitation light source with shutter: Lamps can vary from projector lamps, to Tungsten/Xenon lamps, or LEDs (Spectra-physics model 66921, 1,000 W, Newport Corporation, CA, USA) (see Note 2).

11. Oscilloscope with at least two channels: One is used for viewing a signal from the preparation and one for viewing the stimulus pulses. It is also useful for troubleshooting. (Tektronix TDS 2002, Tektronix, Inc.)

12. Ground wires: All equipment needs to be grounded to a large, low resistance source (see Note 3).

13. Electrical stimulator to stimulate the tissue: Can have either an external pacing device or can pace through computer software with an analog output card (National Instruments Corporation, USA).

14. Antivibration table with thread holes (Technical Manufacturing Corporation) to reduce the amount of mechanically generated noise. Nitrogen gas is used to float the table (see Note 4).

15. Faraday cage: Can be mounted on the table to reduce noise from the surrounding electrical systems, but it is not necessary to achieve low signal to noise recordings (see Note 4).

16. Rack mount system: All electronics, power supplies, and computer should be mounted in a rack system (see Note 5).

17. Defibrillator: Used for terminate tachycardia or fibrillation. (Zoll Defibrillator, model: PD 1200. Zoll medical corporation, Massachusetts, USA).

18. Supports to mount the camera and excitation lamps.

19. BNC cables: To connect amplifier, stimulator, oscilloscope, and optical mapping acquisition box (see Note 6).

20. The optical mapping room: Must have the ability to be completely dark (see Notes 7 and 8).

2.2. Perfusion Setup

1. Tubing: Specific for the pump listed below (Type #16 from Cole-Parmer Instrument Company) or for perfusing the tissue (PE60, Clay Adams, Parsippany, NJ) (see Note 9).

2. Heating coils. (Radnoti Glass Technology, Inc.)

3. A water bath tank (Length × Width × Height $50 \times 40 \times 25$ cm): Set to keep the perfusion temperature between 35 and 37°C. The water bath should have a cover with holes that will enable 500 ml bottles of solution to be suspended in the bath.

4. Heater: Isotemp 2100 heater (Fisher Scientific) and Thermeo Haake DC10 heater (Haake).

5. Pump: Masterflex L/S Digital Modular drive with easy load II (2 channels, Cole-Parmer Instrument Company).

6. Stands to hold heating coils.

7. Tissue chamber: Warm water runs in the outside jacket to keep fluids inside chamber at 37°C. Depth and width of the chamber can vary depending on the size of the preparation. (Radnoti Glass Technology, Inc.)

8. Thermistors: optional, but can be used to measure the temperature of the perfusion and the temperature of the fluid in the tissue chamber. Depending on these temperatures, a TTL pulse can then be sent to the water baths to turn on or off.

9. Modified Tyrode's solution: 123 mM NaCl, 5.4 mM KCl, 22 mM $NaHCO_3$, 0.65 mM NaH_2PO_4, 0.50 mM $MgCl_2$, 5.5 mM glucose, and 2 mM $CaCl_2$, bubbled with 95% O_2–5% CO_2 (12).

10. Cardioplegic solution: 123 mM NaCl, 15 mM KCl, 22 mM $NaHCO_3$, 0.65 mM NaH_2PO_4, 0.50 mM $MgCl_2$, 5.5 mM glucose, and 2 mM $CaCl_2$, bubbled with 95% O_2 to 5% CO_2.

11. Suction: Prevent overflow of fluids in the tissue chamber.

12. Two 10 L glass storage containers: One for the Tyrode's solution, and one to collect the fluids that are removed from the tissue chamber via suction. (Kimble/Kontes, Vineland, New Jersey, USA).

2.3. Preparation Setup
(see Note 10)

1. Cardioplegic solution in a 4 L glass storage container. (Kimble/Kontes, Vineland, New Jersey, USA).

2. Ice bath to keep the cardioplegic solution cold (see Note 11).

3. Pump: Masterflex L/S Digital Modular drive with easy load II (2 channels, Cole-Parmer Instrument Company) and tubing: (Type #16 from Cole-Parmer Instrument Company).

4. Chucks.

5. Clamps, scissors, forceps, and needle driver.

6. Ties and sutures.

7. 95% O_2–5% CO_2 gas mixture bubbled into the cardioplegic solution.

8. Syringe and stop cock.

2.4. Special Chemicals

2.4.1. Excitation/
Contraction Uncouplers
(see Note 12)

1. 2,3-butanedione monoxime: 15–20 mM for perfusion (6). (Sigma-Aldrich, Inc.)

2. Cytochalasin D: 10–40 μM (6, 9). (Sigma-Aldrich, Inc.)

3. Blebbistatin: 5–10 μM for perfusion (7). (Calbiochem, La Jolla, CA, USA.)

2.4.2. Voltage Sensitive
Dyes (see Note 12)

1. Di-4-ANEPPS: 200 μg perfused through the preparation for 15 min (12) (Molecular Probes).

2. PGH-1: 20–30 μL of 2.5 mM stock solution of dye in DMSO plus 16% Pluronic L64 (5) (University of Pittsburgh).

3. Methods
(see Note 13)

3.1. Preparation

1. Make Modified Tyrode's solution and Cardioplegic solution (see Note 14).

2. Turn on the light source at least 30 min before the experiment to stabilize the output.

3. Fill water baths with appropriate amount of water and turn on the heater (see Note 15).

4. Put 400–500 ml of the Tyrode's solution into one of the 500 ml bottles that sit in the water bath and bubble with 95% O_2 to 5% CO_2.

5. Setup tubing for perfusion with the two-channel pump. One line runs from the solution in the water bath bottle to the pump and then to the heating coils to the tissue. Another line from the solution reservoir to the water bath bottle keeps the solution level constant. The tubing used for the two channels needs to be same to keep the same flow through both channels.

6. Turn on the pump to prime the perfusion system with Tyrode's solution and expel any air bubbles from the tubing.

7. Turn on amplifier, oscilloscope and stimulator.

8. Power on the optical mapping acquisition box.

9. Turn on optical mapping computer and launch the acquisition software. Check the computer storage capacity to make sure that there is enough space for recording the data files that will be generated during the experiment.

10. For CCD and CMOS optical mapping cameras, click the "Monitor" button to make sure the optical mapping system is working properly.

3.2. Making the Tissue Preparation

1. Double check to make sure that all components of the setup are working correctly, the perfusion system is connected and ready.

2. Setup surgical area with chucks, needed instruments, suture, tubing, pump, ice, a CO_2/O_2 line, syringe tube, and a plastic cable tie.

3. Fill one container with ice, and place another container with cardioplegic solution within it.

4. Place the remaining cardioplegic solution in an ice bath. Bubble the solution with 95% O_2/5% CO_2.

5. Attach the syringe tube with stop cock to the end of the perfusion line.

6. Prepare the pump and the tubing and prime the system with cardioplegic solution.

7. Excise the heart.

8. Immediately place the heart in the cold cardioplegic solution.

9. Quickly take the heart to the surgical area.

10. Find the aorta and make an incision above the aortic valve in the ascending aorta.

11. With the cardioplegic solution flowing through the perfusion system, insert the perfusion tube with the syringe tube extension.

12. Place the plastic cable tie around the perfusion line in the aorta and close off to prevent backflow of fluid.

13. The flow rate of the pump should be close to its highest levels.

14. Keep the heart in the cold cardioplegic solution while it's being perfused (see Note 16).

15. The blood will now be forced out of the cardiac vessels and replaced with the cold cardioplegic solution. This step could take a few minutes.

16. When all vessels are cleared of blood, turn off the pump.

17. Cut the apex the heart off and pull the syringe tube through.

18. Cut the ventricles away 1 cm below the atrial ventricular ring.

19. If making a left ventricular wedge preparation – continue on to step 20; if making an atrial preparation – jump to step 27.

20. Separate the left ventricle from the right ventricle by cutting down the septum on the right ventricular side. Allow room to tie off vessels.

21. Cut the left ventricular free wall to size. We usually make it large enough to include a papillary muscle.

22. Insert a perfusion line into left anterior descending coronary artery (see Note 17).

23. With a 0-silk suture, tie in the perfusion line at the entrance to the vessel. Make sure that the tie is not too tight to impede perfusion flow. Pull back on the perfusion line such that the beginning of the line is just past the vessel opening.

24. Keep a small amount of perfusion pressure and tie off all vessels.

25. After tying off all visible vessels, increase the perfusion pressure to look for any missed vessels (see Note 18).

26. Once all vessels have been tied off and the preparation can maintain adequate perfusion pressure without any obvious leaks, the preparation is ready for optical mapping.

27. For an atrial preparation, once the atria have been separated from the ventricles, all non cardiac tissue needs to be removed from the preparation. This includes lung and connective tissue.

28. Cut the connective tissue behind the aorta to separate the left and right atria and to expose Bachmann's Bundle.

29. If doing an experiment involving the pulmonary veins, identify each vein, cut the lung off above the point where the cardiac tissue ends, insert a suture at this point.

30. Identify the aorta, and in sections, cut down to the opening to the coronary arteries.

31. Isolate the arterial openings by cutting a shoulder around the vessel opening. Be careful not to cut the vessel. This shoulder is created with just the tissue of the aorta. This shoulder is to enable the perfusion line to be tied in place.

32. There will be two openings for the coronary arteries. Insert one perfusion line into one that feeds the left atrium, and insert one that feeds the right atrium. Once inserted, the line should wrap around towards its corresponding atria.

33. With a 0-silk suture, tie in the perfusion line at the entrance to the vessel, around the shoulder that was created.

34. Make sure that the tie is not too tight to impede perfusion flow. Pull back on the perfusion line such that the beginning of the line is just past the vessel opening.

35. Tie off all open vessels.

36. Increase perfusion pressure to look for any leaks and missed vessels.

37. Once all vessels have been tied off and the preparation can maintain adequate perfusion pressure without any obvious leaks, the preparation is ready for optical mapping.

3.3. Optical Mapping

1. Place the finished preparation in the glass chamber and keep it as flat as possible (see Note 19).

2. Start the Tyrode's perfusion and connect to the perfusion line of the preparation.

3. As the warm modified Tyrode's starts to enter the preparation, the cold cardioplegic solution will be washed out.

4. Suction out this cardioplegic solution from the chamber.

5. While waiting for the preparation to start beating, and the chamber to fill with solution, place the plunge electrodes into the tissue. Two electrodes used for pacing and another pair used for sensing. Place these pairs of electrodes in different sites to avoid stimulus pulse artifacts in the signal seen on the oscilloscope monitor. The actual number of pacing electrodes may differ depending on the protocol.

6. Connect the electrodes to amplifier and simulator.

7. For the CCD or CMOS cameras, click the "Focus Monitor" button from the software window, focus the camera using the room light with the highest resolution, avoid to use the excitation light to minimize dye bleach. Save the focus image with a ruler or grid. For the photodiode camera, use a mirror in the dichroic box that leads to the focusing grid. Place a thin piece of paper with lettering on it to aid in focusing.

8. The preparation should start to beat spontaneously. It is common that after a couple of spontaneous beats, VF develops. Make sure that all of the cardioplegic solution has been removed from the chamber. Defibrillate if necessary and be prepared to pace the preparation.

9. Once sinus rhythm has been achieved and the preparation has strong contractions, inject the voltage sensitive dye into the preparation over a time interval of 5–10 min.

10. As the dye is entering into the tissue and binding to the cell membranes, continue to focus the camera and adjust lighting to make sure the illumination is evenly spread across the recording surface. Make sure to use the correct excitation and emission filters. Save a picture image of the preparation and the recording surface.

11. After a few minutes, the dye should be in the cell membranes – attempt a few recordings to look at signal quality.

12. If necessary – adjust lighting, lens and camera to optimize the signal to noise ratio (see Note 20).

13. Once everything is in place, focused, and adequate signal quality has been achieved, an excitation–contraction uncoupler may be added to the solution.

14. Allow a few minutes for the uncoupler to circulate and the tissue to paralyze.

15. The tissue should stop beating, and all motion artifacts should be removed from the recorded signal.

16. At this point, everything is set to start the protocol. Add more dye if necessary.

4. Notes

1. Older CRT monitors can induce noise into the optical recordings. To reduce noise and scattering of light, LCD flat panel screens should be used for computer screens.

2. The shutter on the excitation light source is important for minimizing the exposure time to avoid dye photobleaching and phototoxicity. The acquisition or stimulation software can control the shutter by TTL pulses.

3. All equipment, lights, power lines, etc. can be a source of noise in the recordings. To help eliminate noise, everything should be grounded.

4. All experiments take place on an antivibration table that is surrounded by a Faraday cage. Both of these are optional, but they help improve the signal to noise ratio as they reduce the amount of mechanically and electrically generated noise with the recorded fluorescence signal.

5. A rack mount system provides a way to organize all electronics and power supplies such that the operator has easy access to all pieces of equipment during the experiment.

6. BNC cables will be used to send the electrical signals from the plunge electrodes in the preparation and the pacing signal to multiple places. It is useful to view these signals on the oscilloscope as well as including them in the data that is recorded. These signals will be sent to the optical mapping box and included in the data file that is created, and to the oscilloscope for viewing.

7. Any light source in the room can result in background noise in the fluorescence signal recordings. For this reason, the room needs to be as dark as possible during optical mapping. All windows need to be covered completely. The illumination from the monitor and the light source needs to be shielded from the camera that is acquiring the signals.

8. Water baths, power supplies, and other pieces of equipment can draw a lot of current. Make sure that the circuits that these pieces are plugged into can handle the amount of current needed to power these electronics. One suggestion is to have only a couple of pieces plugged into one circuit.

9. To make the arterial perfusion lines, the 1.22 mM diameter tube is cut to size. At one end, a 20 Gauge needle with the tip cut off is inserted into the tube. This will enable this perfusion line to be easily connected to any perfusion system. Heat the other end of the tube such that the plastic begins to melt, and creates a shoulder around the opening of the tube. This will help to secure the perfusion line in the artery, and help prevent the line from dislodging during the experiment.

10. The tissue preparation setup area is separate from the optical mapping area and another perfusion setup should be used to create the tissue preparation that will be used in the experiment. All of the equipment used for tissue preparation is similar to what was described in the previous section on the perfusion setup.

11. When making the preparation, it is important to keep it in cold cardioplegic solution.

12. Please see Subheading 1 for a description of excitation/contraction uncouplers and voltage sensitive dyes.

13. Optical mapping involves many pieces of equipment and before performing the actual experiment, it is important to have the basic setup operating without any problems.

14. Can make Tyrode's stock solution without glucose and store it in the refrigerator at 4°C for up to 1 week. The solution PH needs to be adjusted after bubbled with a 95% O_2 and 5% CO_2 gas mixture.

15. Deionized water is recommended to use in the water baths to avoid corrosion.

16. Steps 9–14 are critical steps, and should be performed as quickly as possible.

17. Inserting the perfusion line into the left anterior descending coronary artery can be tricky and you may need a third hand to help secure it in place.

18. In a left ventricular preparation, there will be several vessels that are leaking in the septal wall. Attempt to close off as many as possible.

19. Depending on the size of the chamber, a rubber stopper may be used to prop up the preparation. For an atrial preparation, 0-Silk suture is used to tie down parts of the preparation to make the recording surface as flat as possible (see Fig. 2).

20. For CCD and CMOS cameras from SciMedia, we use a Navitar lens, and the aperture is adjusted to obtain the best image. Usually, a higher saturation of light will result in better signals. But the maximum intensity of the fluorescence should not be over 85% for CCD camera and 95% for the CMOS camera. The type of camera that is used depends on the study aim (see the other chapter 14 within this section for

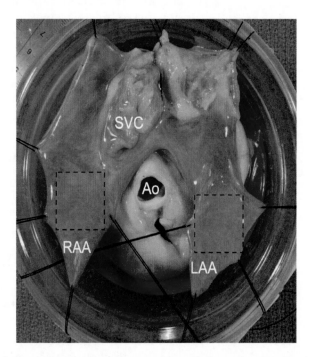

Fig. 2. Perfused atrial preparation, as used in optical mapping experiments. *Black squares* (2 × 2 cm) indicate the approximate recording locations on the right (RA) and *left atrium* (LA) for a photodiode array. CCD and CMOS cameras would be able to record signals from a larger surface area. *SVC* superior vena cava, RAA and LAA, RA and LA appendage, respectively, *Ao* aorta. (Reproduced from ref. 13 with permission from the American Journal of Physiology).

more details). The time and space resolution of the camera can be changed depending on experimental goals. For the CCD camera, the highest time resolution is 1.3 ms and highest space resolution is 384×256 pixels. For CMOS camera, the space resolution is fixed to 100×100 pixels, and the usable time resolution can be high as 0.2 ms or 0.5 ms. Usually, we use 1 ms sampling rate with the CMOS camera and it has a higher signal to noise ratio than the CCD camera. However, the sensitivity of the CMOS camera is less than the CCD camera. Both cameras offer different recording modes and it's worth trying different modes to compare the image quality.

Acknowledgments

This work was supported by AHA Western States Affiliate Beginning Grant-in-Aid 0765177Y (T.H.E. IV).

References

1. Derakhchan, K., Li, D., Courtemanche, M., Smith, B., Brouillette, J., Page, P. L. & Nattel, S. (2001). Method for simultaneous epicardial and endocardial mapping of in vivo canine heart: application to atrial conduction properties and arrhythmia mechanisms. *J Cardiovasc Electrophysiol* 12, 548–55.

2. Everett, T. H. IV., Wilson, E. E., Verheule, S., Guerra, J. M., Foreman, S. & Olgin, J. E. (2006). Structural atrial remodeling alters the substrate and spatiotemporal organization of atrial fibrillation: a comparison in canine models of structural and electrical atrial remodeling. *Am J Physiol Heart Circ Physiol* 291, H2911–23.

3. Schuessler, R. B., Kawamoto, T., Hand, D. E., Mitsuno, M., Bromberg, B. I., Cox, J. L. & Boineau, J. P. (1993). Simultaneous epicardial and endocardial activation sequence mapping in the isolated canine right atrium. *Circulation* 88, 250–63.

4. Ideker, R. E., Smith, W. M., Blanchard, S. M., Reiser, S. L., Simpson, E. V., Wolf, P. D. & Danieley, N. D. (1989). The assumptions of isochronal cardiac mapping. *Pacing Clin Electrophysiol* 12, 456–78.

5. Salama, G., Choi, B. R., Azour, G., Lavasani, M., Tumbev, V., Salzberg, B. M., Patrick, M. J., Ernst, L. A. & Waggoner, A. S. (2005). Properties of new, long-wavelength, voltage-sensitive dyes in the heart. *J Membr Biol* 208, 125–40.

6. Biermann, M., Rubart, M., Moreno, A., Wu, J., Josiah-Durant, A. & Zipes, D. P. (1998). Differential effects of cytochalasin D and 2,3 butanedione monoxime on isometric twitch force and transmembrane action potential in isolated ventricular muscle: implications for optical measurements of cardiac repolarization. *J Cardiovasc Electrophysiol* 9, 1348–57.

7. Fedorov, V. V., Lozinsky, I. T., Sosunov, E. A., Anyukhovsky, E. P., Rosen, M. R., Balke, C. W. & Efimov, I. R. (2007). Application of blebbistatin as an excitation-contraction uncoupler for electrophysiologic study of rat and rabbit hearts. *Heart Rhythm* 4, 619–26.

8. Baker, L. C., Wolk, R., Choi, B. R., Watkins, S., Plan, P., Shah, A. & Salama, G. (2004). Effects of mechanical uncouplers, diacetyl monoxime, and cytochalasin-D on the electrophysiology of perfused mouse hearts. *Am J Physiol Heart Circ Physiol* 287, H1771–9.

9. Hayashi, H., Miyauchi, Y., Chou, C. C., Karagueuzian, H. S., Chen, P. S. & Lin, S. F. (2003). Effects of cytochalasin D on electrical restitution and the dynamics of ventricular fibrillation in isolated rabbit heart. *J Cardiovasc Electrophysiol* 14, 1077–84.

10. Lee, M. H., Lin, S. F., Ohara, T., Omichi, C., Okuyama, Y., Chudin, E., Garfinkel, A., Weiss, J. N., Karagueuzian, H. S. & Chen, P. S. (2001). Effects of diacetyl monoxime and cytochalasin D on ventricular fibrillation in

swine right ventricles. *Am J Physiol Heart Circ Physiol* 280, H2689–96.

11. Wu, J., Biermann, M., Rubart, M. & Zipes, D. P. (1998). Cytochalasin D as excitation-contraction uncoupler for optically mapping action potentials in wedges of ventricular myocardium. *J Cardiovasc Electrophysiol* 9, 1336–47.

12. Everett, T. H. IV., Verheule, S., Wilson, E. E., Foreman, S. & Olgin, J. E. (2004). Left atrial dilatation resulting from chronic mitral regurgitation decreases spatiotemporal organization of atrial fibrillation in left atrium. *Am J Physiol Heart Circ Physiol* 286, H2452–60.

13. Verheule, S., Wilson, E., Banthia, S., Everett, T. H. IV, Shanbhag, S., Sih, H. J. & Olgin, J. (2004). Direction-dependent conduction abnormalities in a canine model of atrial fibrillation due to chronic atrial dilatation. *Am J Physiol Heart Circ Physiol* 287, H634–44.

Part VI

Biomaterials

Chapter 16

Surface Patterning for Generating Defined Nanoscale Matrices

Karen L. Christman and Heather D. Maynard

Abstract

While stem cells in culture have been predominately controlled through the addition of soluble factors to the media, the impact of the extracellular matrix on stem cell renewal and differentiation has recently come to the forefront. In vivo, cells adhere and respond to cues that are on the nanoscale, thus the presentation of extracellular matrix components on this scale is critical to mimicking the in vivo environment. We have developed a highly flexible nanopatterning technique, employing protein and peptide reactive polymers and electron beam lithography, which can be utilized for studying matrix effects on stem cell renewal and differentiation.

Key words: Extracellular matrix, Patterning, Nanotechnology, Stem cells, Differentiation

1. Introduction

The extracellular matrix is known to play a major role in cell differentiation (1–3). In recent years, studies have begun to examine the potential for matrix-directed renewal (4, 5) or differentiation (5–12) of stem cells. Uniformly coated surfaces containing ECM components can be a first step toward deciphering ECM cues that control cell differentiation; however, in vivo, cells are not exposed to this contiguous presentation of adhesion ligands. Chemical patterning of biomolecules at the microscale have been known for some time to modulate cell attachment, growth, and differentiation (13, 14), yet cells recognize and adhere to cell adhesion receptors in vivo that are spatially organized on the nanoscale. Integrins, which are the most common cell adhesion receptor, are approximately 10 nm (15). Integrins also cluster to form focal adhesions, which are typically in the

Randall J. Lee (ed.), *Stem Cells for Myocardial Regeneration: Methods and Protocols*, Methods in Molecular Biology, vol. 660, DOI 10.1007/978-1-60761-705-1_16, © Springer Science+Business Media, LLC 2010

range of 10–200 nm, and necessary for robust adhesion as well as intracellular signaling (16–19). This suggests that presentation of cell adhesion ligands at the nanoscale is necessary for mimicking the ECM and in vivo environment, and in turn, directing cell adhesion and growth or differentiation. Herein, we describe a flexible technology for generating nanopatterned ECM proteins and peptides that will enable the study of ECM effects on stem cell differentiation.

2. Materials

2.1. PEG–ASH

1. 6-arm PEG amide-succinimidylglutarate (ASG), stored at –20°C (The amide linkage and not the hydrolysable ester linkage should be used). The polymer should be stored in aliquots to reduce repeated exposure to moisture.
2. Methanol (reagent grade, Sigma-Aldrich).
3. Clear glass threaded vials (03-338A, Fisher).
4. Thermoset green screw caps with fluoropolymer liner (NC9204470, Fisher).

2.2. PEG–Biotin

1. 8-arm PEG-amine, MW 10,000, stored in aliquots at –20°C (JenKem Technology USA Inc.) (see Note 2).
2. EZ-Link Sulfo-NHS-LC-Biotin, stored in aliquots at 4°C with dessicant (Sulfosuccinimidyl–6–(biotinamido) hexanoate, Pierce).
3. Centriprep centrifugal filter unit 10 kDa cutoff (Millipore).
4. 15 mL conical polypropylene centrifuge tubes (Fisher).
5. Chloroform (reagent grade, Sigma-Aldrich).

2.3. Silicon Wafer Preparation

1. Diamond tip pen (SPI Supplies).
2. Style #6 StudenTek tweezers (SPI Supplies).
3. 1″ wafer carrier trays (Ted Pella).
4. Nanostrip (Fisher).
5. Pyrex Crystallization Dishes (Fisher).
6. Acid waste container.

2.4. PEG Silane Coating for Resisting Protein and Cell Adhesion

1. PEG trimethoxysilane (Prochimia).
2. Anhydrous tetrahydrofuran (THF, Sigma-Aldrich).
3. Glass scintillation vials (Fisher).
4. Easy-squeeze wash bottles (Fisher).
5. Argon gas.

2.5. E-Beam Lithography	1. Tap300 Tapping Mode AFM probes (BudgetSensors).
2.6. Surface Functionalization for PEG–ASG Patterns	1. Desired extracellular matrix protein(s) and/or peptide(s). 2. Sterile phosphate buffered saline (PBS, Gibco).
2.7. Surface Functionalization for PEG–Biotin Patterns	1. Streptavidin (Sigma). 2. Biotinylated extracellular matrix protein(s) and/or peptide(s).

3. Methods

The following technique is based on electron beam (e-beam) crosslinking of poly(ethylene glycol)-based polymers using an e-beam writer. PEG can be crosslinked to hydroxyl bearing surfaces and to itself (20–22), presumably through a radical mechanism. By utilizing functionalized PEGs that can bind to ECM components such as proteins and peptides, the technique can generate nanopatterned matrices for cell studies. Unlike other nanopatterning techniques, this approach is extremely flexible at the nanoscale in terms of pattern shape, size, and spacing. E-beam lithography is, however, hindered by the area that can be patterned and the speed of patterning, thus if larger features are desired, such as those at the microscale, other well-established micropatterning techniques are recommended.

Patterning of two common chemistries for protein and peptide attachment are described. Succinimidyl glutarate reacts rapidly with amines at neutral pH and thus provides a convenient attachment point for any protein or peptide. Many biotinylated proteins and peptides are available and can also be attached to surface biotin patterns through a streptavidin linker. Once these functionalized PEGs are crosslinked into nanopatterns on the surface, noncrosslinked polymer can be easily rinsed away. ECM proteins and peptides can then be immobilized on the surface and utilized for studying cell behavior on different ECM surfaces.

3.1. Preparation of PEG–ASG

1. Remove PEG–ASG from freezer and allow it to equilibrate to room temperature prior to opening, in order to prevent moisture condensation inside the container.

2. Dissolve PEG–ASG (0.04 mg/mL) in methanol inside of a glass vial immediately prior to use. Vortex until solution is clear. Replacement green fluoropolymer lined caps, which are designed for organic solvents, should be used. Long-term storage of this polymer in methanol has not been validated.

3.2. Preparation of PEG–Biotin

1. Remove PEG-amine from freezer and EZ-Link Sulfo-NHS-LC-Biotin from refrigerator, and allow each to equilibrate to room temperature prior to opening.

2. Dissolve PEG-amine (1 mg/mL) in water (see Note 1) inside a 15-mL conical polypropylene tube.

3. Add EZ-Link Sulfo-NHS-LC-Biotin (5 mg/mL) to the tube.

4. Vortex solution for 10 s and incubate at room temperature for 30 min.

5. Place solution in the top of a Centriprep centrifugal unit and turn twist-lock cap to seal the container.

6. Centrifuge the unit at $3,000 \times g$ for 2 min, empty the solution that has flowed through the filter, fill the sample container up to the fill line with water. Repeat this process two more times to ensure removal of any unreacted EZ-Link Sulfo-NHS-LC-Biotin On the last step, do not refill the sample container.

7. Remove the PEG–biotin solution from the sample container and place in a clean 15 mL conical tube. Immerse the tube in liquid nitrogen to freeze, and then lyophilize until dried.

8. PEG–biotin should then be stored in aliquots at –20°C until use.

9. Dissolve PEG–biotin (0.03 mg/mL) in chloroform inside of a glass vial using the replacement fluoropolymer lined green cap. The green caps provide a tighter seal to prevent evaporation and are designed for organic solvents. The solution should be stored at 4°C when not in use for up to 1 week.

3.3. Silicon Wafer Preparation

1. Silicon wafers should be diced in to approximately 1×1 cm chips using a diamond pen. After dicing, chips should be handled only at the corner using the style #6 tweezers and stored in individual wafer trays to prevent scratching.

2. Place diced silicon chips into the bottom of crystallization dish inside a chemical fume hood. Care should be taken so that chips do not overlap and scratch each other. Carefully pour enough Nanostrip to cover the chips and incubate at room temperature for 15 min (see Note 3).

3. Carefully remove the chips, rinse with water by using an easy-squeeze wash bottle for 10 s. Next, place the chips in a new crystallization dish filled with water and keep submerged until further use for up to 24 h.

4. Pour leftover Nanostrip in an appropriate acid waste container. Immerse crystallization dish in a large volume of water and rinse with water to remove any remaining Nanostrip (tap water is fine for glassware rinsing).

5. Silicon chips with only the native oxide are fairly resistant to cell adhesion and may be used in certain studies depending on cell type and length of culture. To further reduce protein and cell adhesion by passivating the surface with a PEG silane, continue to 3.4, otherwise skip to 3.5.

3.4. PEG Silane Coating for Resisting Protein and Cell Adhesion

1. Dissolve 15 mg PEG trimethoxysilane in 1.5 mL anhydrous THF inside a glass scintillation vial.

2. Remove one silicon chip from the water and dry with a stream of air until no visible moisture is left on either side of the chip. Then, immerse the chip in the glass scintillation vial, seal the cap, and sonicate for 1 h. Only one chip per vial should be used to prevent scratching.

3. Remove the chip from the vial, rinse with fresh THF and then water, and dry with a stream of air. Chip should be used the same day for subsequent steps and stored in a glass vial under argon.

3.5. E-Beam Lithography

1. Either a freshly cleaned silicon chip or a freshly silanated silicon chip should be placed on a spin-coater. Pipette approximately 20 μL of PEG solution (either PEG–ASG or PEG–Biotin) onto the chip and spin-coated at 3,000 RPM for 30 s to form a thin film over the entire chip (see Note 4). Then, place the chip on a 70°C preheated hotplate for 30 s to remove any residual solvent.

2. Create a small scratch in one of corner of the chip (Fig. 1) with a diamond tip pen while taking care not to create much debris on the surface.

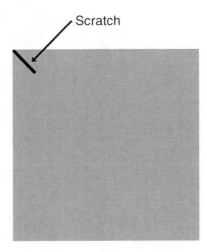

Fig. 1. Schematic demonstrates location of small scratch used for e-beam focusing.

3. Place the chip in an e-beam writer and focus using particles from the scratch (see Note 5). The sample should then be exposed to focused e-beams to crosslink the specific polymer patterns to the surface of the chip. Parameters for each e-beam writer will vary, but example parameters for two different machines are given below (see Note 6):

(a) JC Nabity e-beam lithographic system (Nanometer Pattern Generation System, Ver. 9.0) modified from a JEOL 5910 scanning electron microscope patterning PEG–Biotin: accelerating voltage = 30 kV, beam current: ~4.5 pA, line dose = 3 nC/cm.

(b) Raith50 e-beam writer patterning PEG–ASG: accelerating voltage: 10 kV, beam current: ~10 pA, line dose = 0.3 nC/cm.

4. Remove the chip from the e-beam writer and rinse with either methanol (for PEG–ASG) or chloroform (for PEG–Biotin), followed by water, and then dry with a stream of air. This will remove any noncrosslinked multiarm PEG, leaving only the patterns.

5. Verify pattern size using an atomic force microscope in tapping mode (Fig. 2).

3.6. Surface Functionalization for PEG–ASG Patterns

1. Immerse the PEG–ASG patterned chip in a solution of 1 mg/mL of the desired extracellular matrix protein (i.e., collagen, fibronectin, laminin) or peptide (i.e., RGD) in ~pH 7 PBS for 1 h at room temperature (see Note 7). Then rinse with sterile PBS.

2. Seed desired cell type on the patterns (see Note 8).

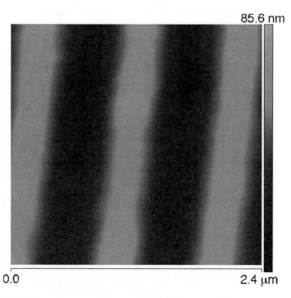

85.6 nm

0.0   2.4 µm

Fig. 2. Atomic force microscope image of PEG–ASG nanopatterns.

3.7. Surface Functionalization for PEG–Biotin Patterns

1. Immerse the PEG–Biotin patterned chip in a solution of 5 μg/mL Streptavidin in sterile PBS for 1 h at room temperature. Then rinse with sterile PBS.

2. Immerse the chip in a solution of 200 μg/mL of the desired extracellular matrix protein (i.e., collagen, fibronectin, laminin) or peptide (i.e., RGD) in sterile ~pH 7 PBS for 1 h at room temperature. Then rinse with sterile PBS. (see Note 9).

3. Seed desired cell type on the patterns.

4. Notes

1. Unless stated otherwise, "water" refers to water that has a resistivity of 18.2 MΩ-cm and total organic content of less than five parts per billion.

2. The technique was originally demonstrated with 8-arm amine-PEG from Nectar, which no longer sells this product. Jenkem sells an 8-arm amine-PEG that should be interchangeable. Other manufacturers such as SunBio also make 4- and 6-arm PEGs, which should also work with this approach.

3. Nanostrip is a stabilized version of Piranha, but care should still be taken to avoid contact with organic materials, metal salts, ammonia, nitric acid, and reducing agents. Piranha solution (3:1, H_2SO_4/30% H_2O_2) may be used an alternative; however, this solution reacts violently with all organic compounds and should be handled with extreme care.

4. Chloroform is very volatile, thus the fastest ramp speed should be used on the spin-coater. If a particular model has a delay after starting, it is advised to first start the spin-coater and then pipette the solution onto the chip immediately before it begins to spin.

5. Typical e-beam lithography used for ablation of resists requires a second focusing step on a burn mark in the resist; however, with this technique a single focusing step on the corner scratch has been shown to be sufficient.

6. We have found that dose is highly dependent on film thickness and example doses are given for the described spin-coating conditions, but because of spin coater variability, this could be altered. Modification of the multiarm PEGs can also require different doses.

7. To save costly protein and/or peptide, it is possible to only place a drop of solution over the pattern, but the chip must be placed in a sealed humid environment to prevent the chip

Fig. 3. Red fluorescent streptavidin attached to PEG–Biotin nanopatterns (*inset*). Nanopatterns are approximately 100 nm separated by 100 nm (AFM image).

from drying out. Lower concentrations may also work adequately but have not currently been validated.

8. If antibiotics are to be used, we have found that sterilizing the chip is unnecessary. If more stringent culture environments are necessary, the chips may be sterilized in 70% ethanol for 1 h and rinse with sterile PBS prior to incubation with any proteins.

9. Fluorescent proteins or antibodies, and fluorescent microscopy can be used to confirm protein attachment if nanopatterns are spaced closely together (Fig. 3).

Acknowledgments

The authors would like to thank Eric Schopf and Kevin Chung in helping to develop these techniques. This work was supported by the NIH NIBIB (R21 EB 005838, HDM), an NIH NHLBI Postodoctoral Fellowship (5F32HL082138-02, KLC), and the NIH Director's New Innovator Award Program, part of the NIH Roadmap for Medical Research, through grant number 1-DP2-OD004309-01 (KLC).

References

1. Glukhova, M. A., and Thiery, J. P. (1993) Fibronectin and integrins in development, *Semin Cancer Biol 4*, 241–249.

2. Czyz, J., and Wobus, A. (2001) Embryonic stem cell differentiation: the role of extracellular factors, *Differentiation 68*, 167–174.

3. Holly, S. P., Larson, M. K., and Parise, L. V. (2000) Multiple roles of integrins in cell motility, *Exp Cell Res 261*, 69–74.

4. Nur, E. K. A., Ahmed, I., Kamal, J., Schindler, M., and Meiners, S. (2006) Three-dimensional nanofibrillar surfaces promote self-renewal in mouse embryonic stem cells, *Stem Cells 24*, 426–433.

5. Gerecht, S., Burdick, J. A., Ferreira, L. S., Townsend, S. A., Langer, R., and Vunjak-Novakovic, G. (2007) Hyaluronic acid hydrogel for controlled self-renewal and differentiation of human embryonic stem cells, *Proc Natl Acad Sci U S A 104*, 11298–11303.

6. Gerecht-Nir, S., Cohen, S., Ziskind, A., and Itskovitz-Eldor, J. (2004) Three-dimensional porous alginate scaffolds provide a conducive environment for generation of well-vascularized embryoid bodies from human embryonic stem cells, *Biotechnol Bioeng 88*, 313–320.

7. Flaim, C. J., Chien, S., and Bhatia, S. N. (2005) An extracellular matrix microarray for probing cellular differentiation, *Nat Methods 2*, 119–125.

8. Hwang, N. S., Varghese, S., Zhang, Z., and Elisseeff, J. (2006) Chondrogenic differentiation of human embryonic stem cell-derived cells in arginine-glycine-aspartate-modified hydrogels, *Tissue Eng 12*, 2695–2706.

9. Goetz, A. K., Scheffler, B., Chen, H. X., Wang, S., Suslov, O., Xiang, H., Brustle, O., Roper, S. N., and Steindler, D. A. (2006) Temporally restricted substrate interactions direct fate and specification of neural precursors derived from embryonic stem cells, *Proc Natl Acad Sci U S A 103*, 11063–11068.

10. Chen, S. S., Revoltella, R. P., Zimmerberg, J., and Margolis, L. (2006) Differentiation of rhesus monkey embryonic stem cells in three-dimensional collagen matrix, *Methods Mol Biol 330*, 431–443.

11. Garreta, E., Genove, E., Borros, S., and Semino, C. E. (2006) Osteogenic differentiation of mouse embryonic stem cells and mouse embryonic fibroblasts in a three-dimensional self-assembling peptide scaffold, *Tissue Eng 12*, 2215–2227.

12. Chen, S. S., Fitzgerald, W., Zimmerberg, J., Kleinman, H. K., and Margolis, L. (2007) Cell-cell and cell-extracellular matrix interactions regulate embryonic stem cell differentiation, *Stem Cells 25*, 553–561.

13. Khademhosseini, A., Langer, R., Borenstein, J., and Vacanti, J. P. (2006) Microscale technologies for tissue engineering and biology, *Proc Natl Acad Sci U S A 103*, 2480–2487.

14. Lim, J. Y., and Donahue, H. J. (2007) Cell sensing and response to micro-and nanostructured surfaces produced by chemical and topographic patterning, *Tissue Eng 13*, 1879–1891.

15. Hynes, R. O. (1992) Integrins – versatility, modulation, and signaling in cell-adhesion, *Cell 69*, 11–25.

16. Burridge, K., Fath, K., Kelly, T., Nuckolls, G., and Turner, C. (1988) Focal adhesions: transmembrane junctions between the extracellular matrix and the cytoskeleton, *Annu Rev Cell Biol 4*, 487–525.

17. Burridge, K., and Chrzanowska-Wodnicka, M. (1996) Focal adhesions, contractility, and signaling, *Annu Rev Cell Dev Biol 12*, 463–518.

18. Shaw, L. M., Messier, J. M., and Mercurio, A. M. (1990) The activation dependent adhesion of macrophages to laminin involves cytoskeletal anchoring and phosphorylation of the alpha 6 beta 1 integrin, *J Cell Biol 110*, 2167–2174.

19. Miyamoto, S., Teramoto, H., Coso, O. A., Gutkind, J. S., Burbelo, P. D., Akiyama, S. K., and Yamada, K. M. (1995) Integrin function: molecular hierarchies of cytoskeletal and signaling molecules, *J Cell Biol 131*, 791–805.

20. Krsko, P., Sukhishvili, S., Mansfield, M., Clancy, R., and Libera, M. (2003) Electron-beam surface-patterned poly(ethylene glycol) microhydrogels, *Langmuir 19*, 5618–5625.

21. Brough, B., Christman, K. L., Wong, T. S., Kolodziej, C. M., Forbes, J. G., Wang, K., Maynard, H. D., and Ho, C.-M. (2007) Surface initiated actin polymerization from top-down manufactured nanopatterns, *Soft Matter 3*, 541–546.

22. Hong, Y., Krsko, P., and Libera, M. (2004) Protein surface patterning using nanoscale PEG hydrogels, *Langmuir 20*, 11123–11126.

INDEX

Randall J. Lee (ed.), *Stem Cells for Myocardial Regeneration: Methods and Protocols*, Methods in Molecular Biology, vol. 660,
DOI 10.1007/978-1-60761-705-1, © Springer Science+Business Media, LLC 2010